SpringerBriefs in Evolutionary Biology

More information about this series at http://www.springer.com/series/10207

Jean Guex

Retrograde Evolution During Major Extinction Crises

 Springer

Jean Guex
Institut des sciences de la Terre
UNIL
Bâtiment Géopolis
Lausanne
Switzerland

ISSN 2192-8134   ISSN 2192-8142   (electronic)
SpringerBriefs in Evolutionary Biology
ISBN 978-3-319-27916-9   ISBN 978-3-319-27917-6   (eBook)
DOI 10.1007/978-3-319-27917-6

Library of Congress Control Number: 2015959946

Springer Cham Heidelberg New York Dordrecht London

Printed on acid-free paper

Springer International Publishing AG Switzerland is part of Springer Science+Business Media
(www.springer.com)

To Catherine, Suzanne, and Raphael

Abstract Most of the evolutionary trends described in the following pages concern more or less gradual geometrical and ornamental transformations occurring over long periods of ecologically stable periods. Such trends are discussed in Chapter 1 of this study.

By contrast, major evolutionary jumps in several invertebrate groups occur during massive extinction periods, which are characterised by the appearance of primitive forms resembling remote ancestors of their immediate progenitors. These forms are defined as atavistic. Homeomorphic species generated during sublethal environmental stress can be separated from the ancestral group by several millions of years. In this paper we present a new theoretical model of retrograde evolutionary changes during sublethal environmental stress and analyse the evolutionary patterns of some planktonic foraminifera, radiolarians, nautiloids, conodonts, corals and ammonoids during major extinction periods. In ecologically stable periods, the transformations of the skeletons are characterised by an increase of shell curvature, corresponding to an increase in the apparent geometrical complexity. During periods of sublethal environmental stress, rapid retrograde evolution occurs in many invertebrates. The evolution of silicoflagellids is discussed as an example of application of artificial stress to modern organisms.

Foreword

For billions of years nature unrolls the script of life, which incessantly amuses scientific minds. The emergence of first prokaryotes and their long dormancy, ensuing sudden explosion of complex forms that appeared first as eukaryotic single cell organisms, subsequent rise in multicellularity and multiple waves of genesis of new phyla, genera and species, constitute a highly convoluted natural history, mechanisms of which remain the main challenge ever faced by human intellect. The path of evolution was turbulent, and often the genesis was interrupted by mass extinction. The responses of life to these mass extinctions are the main theme of this book. The mass extinction events, which the biosphere regularly witnessed, are associated with global catastrophes that suddenly presented living forms with severe stress by narrowing the conditions for survival. This mass extinction, however, not only eliminated species; it triggered a specific response that Jean Guex elegantly described in many papers and in this book as "retrograde evolution." In brief, severe environmental stress instigated the emergence of ancestral forms of many organisms, the forms that have seemingly been extinct for millions of years. In response to cataclysm evolution does not retreat by one step but rather goes all the way down to the basic forms. This is the most fascinating discovery as it not only implies the existence of persistent "genetic memory" but also assumes the presence of an internal switch that activates rapid return to the primeval evolutionary forms. The concept of retrograde evolution and the ideas presented by Jean Guex in this book open a new field in evolutionary biology: the field of "genetic memory" and rapid regulation of evolutionary trends in response to environmental challenges. What are the intrinsic mechanisms, how is the genetic memory preserved, why are primitive forms more resilient and what forces rapid retreat? All this remains an open problem. This conundrum is also a challenge, and this book is made to excite the intellectual capacities of the reader and inspire further advances in our evolutionary thoughts.

Manchester
October 10, 2015

Alexei Verkhratsky

Acknowledgements

My warmest thanks go to Alex Verkhratsky for his very careful editorial revision of this little book. I also thank my old friends for our long-lasting discussions: Spela Gorican, Elizabeth Carter, Ursel Haeusler, Annachiara Bartolini, Luis O'Dogherty, Paulian Dumitrica, Arthur Escher, Federico Galster, Daniel Truempy, Dave Taylor, Aymon Baud, François Reuse, John Hamal Hubbard and Peter Baumgartner. I also wish to thank my friends who disappeared recently: Milos Rakus, Jean Marcoux, Jean Gabilly, Rudolf Truempy and Renaud du Dresnay. My thanks also go to Laurent Keller and Philip McGough for interesting suggestions.

Contents

Chapter 1
Evolutionary Trends During Periods of Relative Ecological Equilibrium

1.1 Introduction

The global crises which punctuate Earth History and are responsible for the classical extinctions that took place—for example, during the Permian Triassic or at the Cretaceous Tertiary boundary—are framing periods of relative ecological equilibrium during which gradual, slow evolutionary trends lasting millions of years develop.

These trends often start with "simple" organisms and end with much more "advanced" ones. These rather obvious trends have been widely studied during recent decades at the expense of researching the evolutionary reactions of the living organisms during the crisis.

The goal of the present book is to analyse the reversals of the most common evolutionary trends observed in the fossil record of some selected invertebrates (ammonoids, radiolarian, foraminifera and conodonts) during periods of ecological disequilibrium and extinctions. It is therefore necessary to describe first the basic trends observed during periods of ecological dynamic equilibrium.

Then we will describe the reversals affecting the organisms which survived major extinctions generated by ecosystem collapses due to both extrinsic (physical and/or chemical) and intrinsic (biotic) stresses.

Palaeontologists generally use the term "evolutionary trend" to describe the oriented morphological transformations occurring in stratigraphic sequences of one particular species or in phyletic series of closely related species. In some cases, trends seem to be gradual and are used as a biochronological clock for stratigraphic correlations (Peybernes et al. 1997; Hottinger 1981; Less and Kovacs 1996). However, in most cases, they appear as discrete sequences of closely related species belonging to a single lineage showing an oriented morphological variation.

The phyletic increase in body size is the most frequently quoted evolutionary trend. It is known as Cope's Rule, named after the American vertebrate palaeontologist who first observed it in the nineteenth century (Cope 1896). The most famous case, illustrated in many palaeontological textbooks, is the controversial evolution of the horse.

© Springer International Publishing Switzerland 2016
J. Guex, *Retrograde Evolution During Major Extinction Crises*, SpringerBriefs
in Evolutionary Biology, DOI 10.1007/978-3-319-27917-6_1

1.2   Evolutionary Trends in Ammonites

1.2.1   Initial Trends

Today it is banal to observe that the most frequent long-lasting evolutionary trend observed in Mesozoic ammonites is that where the ancestral group has an opened umbilicus (evolutes form) and the derived group has a closed umbilicus (involute form). This tendency was recognized more than hundred years ago (Hyatt 1889) and the same trend characterizes the Devonian bactritids from the very beginning of their history (Erben 1966) (Fig. 1.1). When completely accomplished, the increasing spiralization of initially evolute shells gives rise to lenticular (oxy-cones) or spherical (sphaerocones) forms (Fig. 1.2). In the following discussion

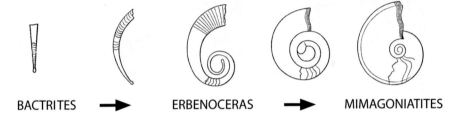

BACTRITES   ➡   ERBENOCERAS   ➡   MIMAGONIATITES

Fig. 1.1 First coiling observed in the Devonian ancestors of ammonoids (Redrawn and modified from Erben 1966. From Guex (2001))

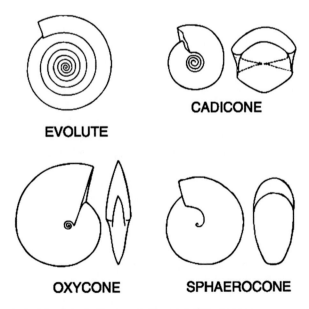

CADICONE

EVOLUTE

OXYCONE   SPHAEROCONE

Fig. 1.2 Some typical involute shells generated by an evolute ancestor

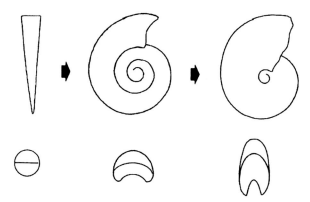

Fig. 1.3 Increasing curvature leading to involute geometry

we consider that the spiralization increase corresponds to an elevated apparent geometrical complexity of the shells because it is the consequence of a double increase of the curvature of these shells: first the spiralization of the originally straight tube and second the increase in involution (tight coiling) (Fig. 1.3). Similar trends are also seen in nautiloids (Sobolev 1994) and certain gastropods (Runnegar 1987).

Generally speaking such increases of the shell curvature are related to the surface increase "rule" observed in so many fossil lineages and described as the "generalized Cope's rule" by Guex (2001, 2003).

1.2.2   Size and Involution Increase in the Acrochordiceratidae

Monnet et al. (2012a, b) analysed the evolutionary trends of the family Acrochordiceratidae Arthaber from the Early to Middle Triassic (251–228 Ma). The study was based on very large samples of this ammonoid family which were obtained from strata in north-west Nevada and north-east British Columbia. They enable quantitative and statistical analyses of its morphological evolutionary trends and demonstrate that the monophyletic clade Acrochordiceratidae underwent the classical evolute to involute evolutionary trend associated with an increase in its adult shell size (diameter) and an increase in the complexity of its suture line. This trend in ammonoid geometry is a beautiful illustration of Cope's rule with an accommodation of the increase of the shell diameter and involution (Fig. 1.4).

The recurrent character of this kind of trend was first discussed in the early 1940s to explain the multitude of heterochronous homeomorphies observed within this group (Schindewolf 1940; Haas 1942). Some ammonite lineages also show a broad trend towards increased sinuosity of the growth lines and, on a large timescale, this group shows an overall increase in suture line complexity.

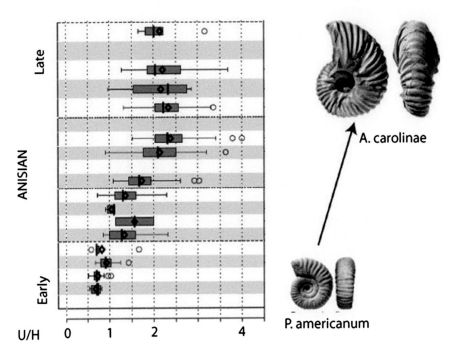

Fig. 1.4 Increasing size and involution (U/H=umbilicus vs. height) in the Acrochordiceratinae during the Middle Triassic (From Monnet et al. 2012a, b)

A sharp description of the iterative evolution of the ammonoids has been produced by Haas (1942, p 643): «…not only types and groups of types reiterate themselves in the history of ammonites, but also certain evolutionary cycles, each proceeding along definite anagenetic trends, e.g., from evolute, sturdy forms with coarse costation to more involute and discoidal ones with a finer and more sigmoidal ribbing»

1.2.3   Paedomorphic Genesis of a Sphaerocone Ammonite

During the middle and late Early Sinemurian, some microderoceratids start to develop a tendency to form globose geometry in the juvenile, prefigurating the proterogenetic (paedomorphic) transformation of the evolute spinose Microderoceras (Early Sinemurian Eoderoceratid) into a sphaerocone Liparoceras (Early Pliensbachian) via an intermediate group called Tetraspidoceras (Late Sinemurian) (Fig. 1.5).

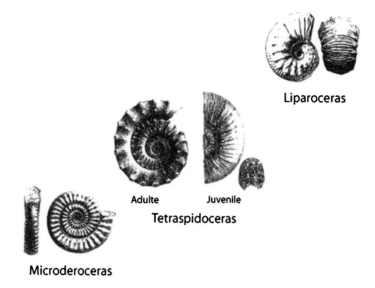

Liparoceras

Adulte   Juvenile
Tetraspidoceras

Microderoceras

Fig. 1.5 Paedomorphic genesis of a sphaerocone *Liparoceras* within *Tetraspidoceras* starting from an evolute spinose ancestor *Microderoceras* (not to scale)

1.2.4   Rates of Changes in the Involution of Some Ammonites

Following the major End Triassic Extinction we can follow the evolution of several genera deriving from the smooth and very simple *Psiloceras* of the Lower Hettangian: *Kammerkaroceras* via *Discamphiceras* (Discamphiceratinae, Psilocerataceae), *Angulaticeras* via *Saxoceras* and *Kammerkarites (Schlotheimiidae, Psilocerataceae), Pseudaetomoceras* via *Caloceras* and *Alsatites (Arietitidae, Arietitaceae), Badouxia* via *Caloceras* and *Sunrisites*. Thanks to new geochronological data produced by Guex et al. (2012a), the rate of change in the involution can be quantified and expressed by the variation of the ratio U/D (=umbilicus vs. diameter) in some typical phylogenetical lineages mentioned above (Fig. 1.6).

1.3   Increasing Involution in Foraminifera

The increase in involution equally affects many unicellular organisms such as planktonic and benthic foraminifera (e.g. the appearance of *Orbulina*) at various stages of their development (Septfontaine 1988; Adams 1983; Blow 1956; Cifelli 1969). Certain benthic foraminifera exhibit an increase in lateral elongation (Hottinger and Drobne 1988) that geometrically corresponds to the development of

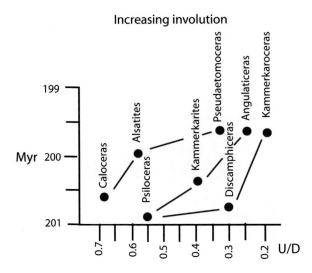

Fig. 1.6 Increasing involution in Lower Jurassic main subfamilies (Guex et al. 2012a)

cadicone coiling in ammonites (Fig. 1.2). In other groups of microfossils discussed below in more detail, such as nassellarian Radiolaria, a similar phenomenon is observed, namely increased sphericity and reduction in the number of segments (Sanfilippo and Riedel 1970). This results in the development of cryptocephalic and cryptothoracic forms (Dumitrica 1970). Similarly, within silicoflagellates (Chrysophytes) discussed in Chap. 8, we find that globular shells such as *Cannopilus* have evolved from simple spicular forms (Guex 1993). Figure 1.7a–d represents various modes of increasing shell curvature through the evolution of foraminifera. Note that (1) these modes of transformation are sometimes associated with an increase in size and (2) the increased involution can occur at any stage of the development but it is more frequently peramorphic than paedomorphic.

1.4   Morpho-Functional Interpretations

Palaeontologists who described the above-mentioned trends have frequently proposed ad hoc adaptive and morpho-functional explanations. Observed increase in involution among ammonites is generally explained in terms of optimal use of the shell material, shell strength and/or improved streamlining (Raup 1967). However, the geometrical antinomy between involute lenticular shells and spherical shells means that the trend towards increasing involution is not uniquely the result of an optimization mechanism for shelly material usage. This is in fact because involute lenticular forms, which are abundant in the fossil record, are far from an optimal geometry from this point of view. Other authors suggest that increased shell

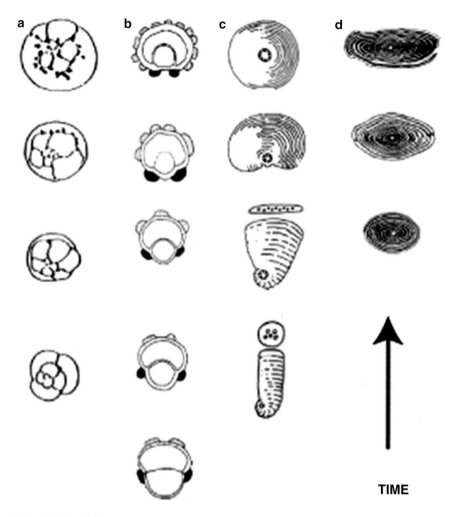

Fig. 1.7 Foraminifera: (**a**) Increasing involution in the *Orbulina* lineage (planktonic foraminifera (Blow 1956; Cifelli 1969)). (**b**) Cryptoproloculinization in *Lepidocyclina* (benthic foraminifera (Adams 1983)). (**c**) Increasing involution in lituolids (benthic foraminifera (Septfontaine 1988)). (**d**) Benthic foraminifera elongation in alveolinids (benthic foraminifera (Hottinger and Drobne 1988)). From Guex (2003)

involution and greater complexity of sutures increase the shell's resistance to hydrostatic pressure (Hassan et al. 2002). Increased surface area of benthic foraminifera is usually interpreted as aiding oxygen exchange. Similarly, elongation of the test is sometimes explained as an optimization of the animal's motility within unconsolidated sediment. As for size increase, the most frequent explanation is morphofunctional (better resistance to predators) or invokes the famous mantra "*nowhere but up!*" (Stanley 1973; McKinney 1990).

1.5 Decoupling of Volume, Surface and Linear Dimension

The above ad hoc explanations are hardly satisfactory because the trends discussed in this introduction are observed in very diverse phyla, including planktonic, nectonic, benthic and burrowing organisms. Moreover, it is also well known that continuous size increase usually leads to gigantism, which can prove to be fatal in a more or less short time. The allometries observed during the geometrical/morphological evolution of shelly invertebrates show that size (i.e. diameter or length: see above), volume and surface can vary independently.

Within ammonites, an increase in volumetric size, which is not accompanied by an increase in linear size (i.e. the diameter; note that the body chamber's length is often unknown for preservational reasons), will result in an increase of involution. Similarly, a decrease in linear size which is not accompanied by a decrease in volume will also lead to a drastic increase of involution. Such a process certainly accounts for the geometry of the lower Triassic small cryptogenic ammonites such as the spherical Isculitids deriving from the serpenticone Columbitids. We also note that an increase in the mantle's surface area, if not compensated by a simultaneous increase in volume of the animal, results in an increase in suture complexity and/or flexuosity of growth lines at the aperture. On the other hand, a decrease of the volume not compensated by a decrease in the mantle surface can explain the small juvenile bulges observed in the inner whorls of the primitive Psiloceratids such as the "Knötchenstadium" of *P. spelae* (syn. *P. spelae tirolicum*).

Another interesting by-product of the above-described morphogenetic rules is the "stop and go" growth of the ammonites followed by an oblique reorientation of the growth lines (Guex 1967, pp 328 329). This is obviously due to the fact that the growth stop of the shell secretion is followed by a delayed restart of the soft parts' growth, generating a rotation of the growth lines.

1.6 Radiolarian Evolutionary Trends

1.6.1 Introduction

Radiolarians are Cambrian to recent holoplanktonic marine protists with morphologically very diverse siliceous skeletons. In the Mesozoic, two main groups are differentiated: nassellarians (mostly conical, composed of one or more consecutive segments) and spumellarians s.l. (generally spherical, composed of one or more concentrical shells). Following the extensive radiolarian research over the past decades, it is now possible to trace the development of some Mesozoic radiolarians through time and to reliably reconstruct several phyletic lineages. In this section we analyse some lineages with well-marked trends in skeletal development (and compare these trends with those observed in other marine organisms). The most usual geometrical transformations occurring in radiolarians are characterized by an increase of the surface of the shell. In several nassellarians, we observe a progressive inflation of the test, leading to a spherization. In some groups such a trend leads to cryptocephalization (Dumitrica 1970), a phenomenon analogous to the orbulinization of some Tertiary foraminifers. The development of a terminal tube or of apertural arches is also frequent in the nassellarians. In many respects, the transformations observed in the spumellarians are related to the same kind of geometrical modifications. For example, the addition of an arch to a polar spine in *Baumgartneria*, of a ring in the Saturnalids, of a button or spine also in the ring of *Aurisaturnalis*, are clear examples of an increase of the shell surface through time. Several examples of evolutionary reversals are given in Chap. 4 where the radiolarian is discussed. For a comprehensive and general discussion on radiolarian systematics and terminology the reader is referred to the monograph by De Wever et al. (2001).

1.6.2 *Peramorphic Trends and Isometric Size Increase*

Isometric size increase (i.e. without change in shape) is rare in radiolarians. Good examples are found in *Spongostichomitra* (Fig. 1.8) and *Obeliscoites*. *Spongostichomitra* has a conical to cylindrical test with very thick and unconstricted spongy wall.

Primitive forms have a small number of chambers, which are not visible in evolved species. Broken specimens show that the segmental divisions are lacking or poorly developed. This group shows an exceptional size increase between the Albian-Cenomanian transition.

Fig. 1.8 Isometric size increase in *Spongostichomitra* from the Albian-Cenomanian (scale bar 100 μm). Guex et al. (2012b)

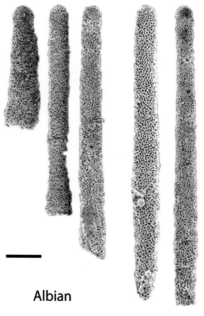

Albian

Cenomanian

1.6.3 Peramorphic Trends and Surface Increase: Spherization of the Test

The most frequent and best described trend is spherization of the test where the increase of volume is not or only weakly coupled with an increase of the length of the shell (Riedel and Sanfilippo 1981; Sanfilippo et al. 1985; Carter and Guex 1999). A similar trend was first described by Sanfilippo and Riedel (1970, 1982) in the Cenozoic Thecotylidae where it is accompanied by a reduction of the cephalothorax. It is also well illustrated in *Palinandromeda* (Fig. 1.9) and *Mirifusus* (Fig. 1.10).

Palinandromeda is a broadly conical amphipyndacid with 5–7 segments and a large basal aperture (Fig. 1.9). The cephalis is dome shaped, poreless and usually has an apical horn. The thorax is small, cylindrical and directly joined to the cephalis. The abdominal and postabdominal segments are trapezoidal or bell shaped with large pores increasing in size distally. The evolution of this group is characterized

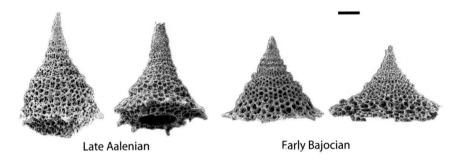

Late Aalenian   Farly Bajocian

Fig. 1.9 Evolution of *Palinandromeda* from the Late Aalenian-Early Bajocian. Scale bar 100 μm. From Bartolini in Guex et al. (2012b)

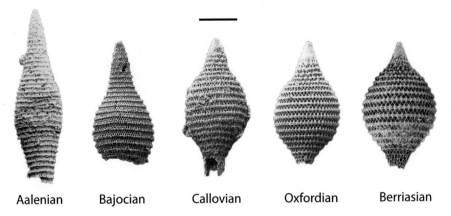

Aalenian   Bajocian   Callovian   Oxfordian   Berriasian

Fig. 1.10 Spherization of the test in *Mirifusus* from the Aalenian-Berriasian. Scale bar 100 μm. Guex et al. (2012b)

by an increase in the flatness of the abdominal segment, developing a planar basal surface. This evolutionary transformation is an extreme exaggeration of the spherization trend whereby the radius of the sphere becomes so large that this nassellarian becomes widely bell shaped and flat during its phylogeny.

The test of *Mirifusus* (Fig. 1.10) is spindle shaped during the Aalenian except in the earliest forms such as *Mirifusus proavus*, and is probably derived from *Parvicingula*-like nassellarians, where it is still slightly conical with the lower part of the spindle poorly developed. The proportions and shape of the conical proximal and inflated median part of the test is highly variable but during its evolution, *Mirifusus* shows a strong increase in sphericity and many species of this genus tend to become spindle shaped and even can develop a terminal tube. It is interesting to note that the duration of the main evolutionary transformation from *M. proavus* to *Mirifusus minor* lasts roughly 20 million years (late Aalenian to late Oxfordian). Then *M. minor* remains practically unchanged for more than 20 million years until the late Hauterivian, when it becomes extinct. This clearly means that the end form can survive successfully for a long period of time if environmental conditions remain favourable.

Another important case of spherization occurs in the williriedellids (Fig. 1.11), which show an extreme increase of the involution, analogous to that observed in Neogene orbulinid foraminifers (Kennett and Srinivasan 1983; Bolli and Saunders 1985; Guex 1992, 1993, 2003). This phenomenon was described as cryptocephalization by Dumitrica (1970).

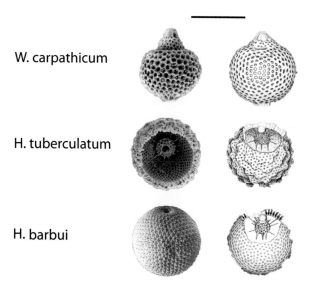

W. carpathicum

H. tuberculatum

H. barbui

Fig. 1.11 Cryptocephalisation within the family Williriedelidae (*Williriedelum carpathicum, Holocryptocranium tuberculatum* and *H. barbui*). Scale bar 100 μm. From Dumitrica 1970 in Guex et al. (2012b)

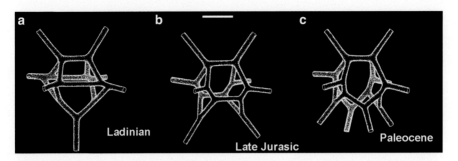

Fig. 1.12 Duplications of the antapical spine in Centrocubus between the Ladinian and Paleocene. Scale bar 100 μm. From Dumitrica in Guex et al. (2012b)

1.6.4 Paedomorphic Trends in Centrocubidae

Proterogenesis occurs sometimes in nassellarians and spumellarians s.l., especially in the Pyloniaceae and Centrocubidae where new structures first develop in the earliest ontogenetic stage. The initial antapical spicule of Centrocubids frequently evolves by doubling or quadrupling of the antapical spine, leading to entirely new modes of growth in subsequent descendants and to forms which are cryptogenic with regard to their ancestors. The evolution of the microsphere of some selected *Centrocubus* is illustrated in Fig. 1.12.

1.7 Cope's Rule, Surface Increase and Apparent Complexity

Before going further we should briefly discuss the concept of complexity applied to living organisms which is used, most of the time, from a purely intuitive point of view (Fig. 1.13) (see McShea 1991; Lloyd 2001).

In some cases it is possible to assign a numerical value to the apparent complexity of a particular character. For example, an ammonite suture line with a high value of its fractal dimension will look more "complex" than a suture line with a low fractal dimension (Guex 1981). The global curvature of a shell provides also a measure of its geometrical complexity, a straight shell looking more "simple" than one which is highly contorted or tightly coiled (Figs. 1.1 and 1.2). It should also be clear that two concepts should be distinguished from the very beginning, with the concept of geometrical complexity coming first and the idea of ornamental complexity being secondary. Sometimes the two variables covary during the ontogeny and sometimes they do not.

This distinction is important when analysing the covariation under the light of the very frequent evolutionary trend of phyletic size increase.

The size increase, however, is not a truly general evolutionary rule because many lineages do not increase in size during their evolution and numerous cases of size decrease are known in the fossil record. For example, a remarkable case of drastic

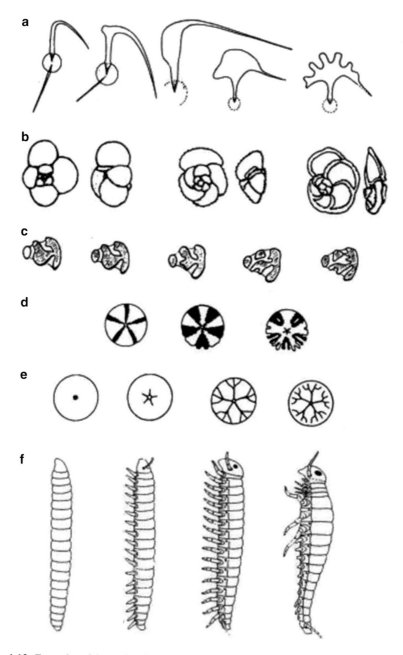

Fig. 1.13 Examples of the surface increase rule generating an apparent complexity increase. All figures simplified and not to scale (from Guex 1992). (**a**) *Oertlisponginae* (Dumitrica 1970). (**b**) Development of the carina in planktonic foraminifera (Cifelli 1969). (**c**) Complexification of the teeth in the *Piezodus-Prolagus* lineage (Hürtzeler 1962). (**d**) Development of the marginal notches in *Rotulidae* (Durham 1966). (**e**) Development of the actinal groves in *Proscutellidae* (Durham 1966). (**f**) Transitions from annelids to insects via onychophorids and myriapods (Raff and Kaufman 1983). From Guex (1992)

size decrease has been observed in planktonic foraminifera below the Cretaceous-Tertiary boundary which is certainly due to the environmental stress generated by the giant volcanism of the Deccan traps (see Courtillot 1999). These size fluctuations and trend reversals are usually dependent on environmental variations and increasing size is mainly observed at the beginning of phyletic lineages.

The allometries observed during the geometrical/morphological evolution of shelly invertebrates show that size (i.e. diameter or length), volume and surface can vary independently. Within ammonites, an increase in volumetric size, which is not accompanied by an increase in linear size (e.g. the diameter), will result in an increase of involution (Figs. 1.1 and 1.2). Similarly, a decrease in linear size which is not accompanied by a decrease in volume will also lead to an increase of involution, leading to subspherical forms. It is obvious that an increase in the mantle's surface area, if not compensated by a simultaneous increase in volume of the animal, results in an increase in sutural complexity and/or flexuosity of growth lines at the aperture in ammonoids, etc. An increase in the involution and in the apparent complexity of internal structures observed in tests of many foraminifera, as well as the increasing complexity of some ornamental traits or of geometrical features that are observed within many marine shelly invertebrates is indirectly related to this generalized Cope's rule (rule of surface increase). To summarize we can say that from a geometrical point of view, all the trends described above are the result of an increase of the surface of the membrane responsible for the biomineralization. The ornaments themselves are subordinated to the general geometry of the shell, specially the details of the local curvature of its secreting membrane, as shown below (Chap. 5, Fig. 5.10).

1.8   Technical Remark About "Full House" (Gould 1996)

In 1996, Gould published a famous book, "Full House", in which he tried to demonstrate the absence of a global increase in complexity during the evolution of living organisms. His main argument is that the most abundant organisms present on the Earth, the bacteria, did not really evolve morphologically during the last 3 billion years. In this book, Gould uses quantitative arguments like size variations observed in the Cretaceous and Cainozoic foraminifera and the fractal dimension of the Ceratites suture lines during the late Palaeozoic and early Mesozoic. We will briefly examine those two quantitative arguments.

His figure 25 (1996, p 160) shows the size variations of planktonic foraminifera during the Late Cretaceous and the Cainozoic. These diagrams are supposed to demonstrate that the rule of size increase (Cope's rule) is meaningless because it is often represented by zigzag variations, as is the case in the planktonic foraminifera observed during that period. We consider that Gould's argument is invalid because size decreases are always related to environmental stress, a phenomenon known since the beginning of the twentieth century (Shimer 1908; see also Mancini 1978). Gould's diagrams (his Fig. 25) just illustrate in a perfect manner that the size reductions occurred during the KT boundary, Late Eocene and Late Oligocene, which are periods of high environmental stress responsible for more or less pronounced extinctions (see Zachos et al. (2001) and Schmidt et al. (2004) for recent quantitative data).

His second quantitative argument concerns the fractal dimension of some ammonoid suture lines during the late Palaeozoic and Triassic times (loc.cit Fig. 35, p. 210). The measurements representing these relationships, as constructed by Gould, are distributed in a completely chaotic way and are supposed to demonstrate that there are no relationships between sutural complexity and time during that period.

The problem is that Gould overlooked the fact that he should have connected the dots representing the various measurements following two criteria. First the ontogeny: small juvenile specimens have a suture line, which always looks less complex than an adult one. And secondly the phylogeny within each separate lineage: the suture line of most ceratitids becomes more complex during the evolution. This is well illustrated in our Fig. 5.3 which shows that the ceratitic suture line of the ancestor of the highly complex phylloceratids has a fractal dimension of about 1.2 when the resulting advanced *Phylloceras* has a suture line dimension of about 1.6 (see Guex 1981). The psiloceratid *Neophyllites* generated during the TJB extinction period also shows a drastic reduction of its sutural complexity.

Chapter 2
Sublethal Environmental Stress

2.1 The Geological Record

The concept of sublethal environmental stress denotes specific conditions that are critical to the survival or normal development of living organisms. The most common cases of environmental perturbations include pollution or poisoning by chemicals such as toxic gases, nutriment shortage, large scale sea level falls, major climatic changes, hydric stress, acid rains, marine anoxies, etc. The origins of environmental stress can easily be identified and quite naturally the particular kind of stress can be hostile to some organisms and favourable to others.

The role of natural environmental stress on development and evolution is widely accepted (Badyaev 2005; Bijlsma and Loeschcke 2005; Nevo 2011; Kishony and Leibler 2003; Hoffman and Parsons 1991; Rutherford and Lindquist 1998; Hangartner et al. 2011; Cabej 2011; Jablonka and Raz 2009; Jablonka 2013; Schlichtling and Pigliucci 1998). The nature of sublethal environmental stress that occurred during major extinctions such as Permian-Triassic (PT), Triassic-Jurassic (TJ), Pliensbachian-Toarcian (Pl-To) and Cretaceous-Tertiary (KT) is currently intensively investigated by means of geochemical and geochronological studies (see, for example, the data in Figs. 2.3 and 2.5.). However most problems related to these exceptional situations belong to the realm of palaeontology and most, if not all, of such palaeontological studies are dedicated to a census of the biodiversity variations (counting how many groups survive, how many disappear, etc.) rather than to the understanding of the basic phenotypic and epigenetic variations induced by major environmental perturbations. The modes of evolution during major extinction events, the characteristics of the organisms surviving major crises and what types of transformations have affected them have not been explored. The main goal of the present study is precisely to analyse in detail the transformations of some invertebrates during major extinction periods.

© Springer International Publishing Switzerland 2016   17
J. Guex, *Retrograde Evolution During Major Extinction Crises*, SpringerBriefs
in Evolutionary Biology, DOI 10.1007/978-3-319-27917-6_2

The geological record of environmental stress is naturally poor because local pollution, nutriment levels, paleotemperatures, etc. can hardly be deduced neither from sediments nor from fossils. The primary indication and proof of the sporadic sublethal environmental stress are the extinction periods themselves, which are easily recognized from the presence of major faunal turnovers reflecting the disappearance of entire phyla and their replacement by new ones (Sepkoski 1978; Benton 1995).

The classical explanations of the origins of extinctions include extraterrestrial impacts, marine regressions, climatic changes or anoxic events (Hallam and Wignal 1997; Hart 1996; Sepkoski 1978; Benton 1995; see also review by Courtillot and Gaudemer 1996). The theory of giant volcanism (Courtillot 1999) has the most general explanatory power, considering enormous potential consequences to the chemistry of seawater, global climate and temperatures. There is an almost perfect correlation between the major extinctions and periods of volcanism (Fig. 2.1, from Courtillot loc.cit.). Virtually all major extinctions are related to the major ecological instability generated by giant volcanism, i.e. climatic changes, atmospheric poisoning by sulfur gases, as well as by darkening generated by fine particles and aerosols inducing major coolings and fall in the sea level (Fig. 2.2). In extremes, for example, at the Permian-Triassic boundary, the only organisms found in crisis sediments are microbialites and fern or fungi spores. Often these sediments do not contain any fossils and hence are known as barren intervals. The after extinction recovery is generally characterized by the explosion of simple and primitive life forms, which can be divided into two major groups:

1. Persistent opportunistic simple forms such as bacteria, fungi, ferns and some simple forms or other microfossils.
2. Primitive-looking forms derived from their immediate ancestors by retrograde evolution (a phenomenon that has been described as proteromorphosis), often associated with a reduction in body size. These organisms are not Lazarus taxa because their absence in sediments older than the ones where they are found is fully reproducible worldwide and the duration of the intervals of time in which they are totally absent can last several millions or several tens of millions of years. In other words, a "Lazarus" explanation of such organisms cannot be reconciled with the geological record.

The present paper analyses how different groups (Foraminifera, Radiolaria, Ammonoids, Nautiloids, Corals, Conodonts and Silicoflagellids) of invertebrates (except conodonts) survive the sublethal biotic crises during extinction episodes and elaborates previously published works (Guex 1992, 1993, 2001; 2006). This study also aims at developing a model explaining the heterochronous repetition of similar evolutionary lineages without invoking repetitions of identical environmental conditions.

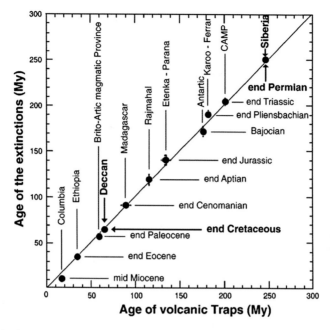

Fig. 2.1 Bivariate graph showing the correspondence between the principal mass extinctions and their geochronometrical age From Courtillot (1999), modified

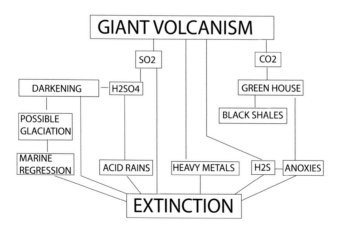

Fig. 2.2 Summary of the most obvious relationships between giant volcanism and major extinctions

2.2 Geochemistry, Geochronology and Volcanic Stress

2.2.1 Introduction

Two major extinctions generated by giant volcanism have been recently studied in detail by the author and his colleagues: the Triassic-Jurassic and the Pliensbachian-Toarcian boundaries. These two major geological events will be discussed in the following sections in light of recent geochemical and geochronological data and will illustrate the environmental framework of the retrograde evolutions observed during these events and discussed in Chaps. 4–6.

2.2.2 A Precise Timing of the End Triassic Extinction

The End Triassic Extinction has long been suspected to be related to the onset of the Central Atlantic Magmatic Province (CAMP) volcanism but it is only recently that U-Pb ages measured on zircons have allowed a precise correlation with the relative timescale based on the evolution of the ammonites. That correlation has been established based on very detailed stratigraphic research done in the American Cordilleras (Northern Peru and Nevada, USA) and on the discovery of ash beds deposited in the same levels as age diagnostic ammonites (Guex 1995; Schoene et al. 2010; Guex et al. 2012a). These discoveries allowed us to propose an original model explaining the precise timing of the End Triassic extinction (ETE) (Fig. 2.3).

One popular model to explain the ETE catastrophic event invokes super-greenhouse conditions due to extreme atmospheric CO_2 concentrations (McElwain et al. 1999; Schaller et al. 2011). This enrichment is often interpreted as degassing of magmatic CO_2 from huge volcanic basalt provinces (e.g. Sobolev et al. 2011 for the Permian-Triassic crisis) and/or from the degassing of carbonaceous or organic rich sediments (e.g. Svensen et al. 2009).

The second scenario invokes a short period of global icehouse conditions caused by degassing of huge volumes of volcanic SO_2, atmospheric poisoning, cooling and eustatic regression coeval with the main extinction (ETE) but probably older than the main basalt emissions. As mentioned above, this model uses the same arguments as those given in Sect. 2.2.3 for the Late Pliensbachian cooling event.

Although both hypotheses are compatible with massive volcanic degassing related to the emission of large volumes of flood basalts, they must also be able to explain the palaeontological record in complete stratigraphic sections that displays decoupling between the (marine) ammonoids vs. (terrestrial) plant extinctions (Guex et al. 2012a). Correlating the sedimentary and the fossil record with carbon and oxygen isotope variations and sea level changes from the T-J and Pl-To boundaries indicates that both boundaries are related to a regressive event followed by major sea level rise (Guex et al. 2001, 2004, 2012a).

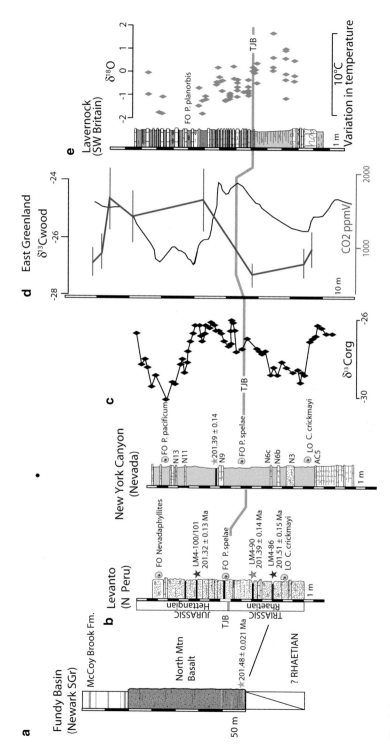

Fig. 2.3 (**a–c**) Numerical ages of the base of the North Mountain basalt and of the TJB in Peru and Nevada. δ¹³C curve at New York Canyon (Nevada). From Schoene et al. (2010), Guex et al. (2004) and Bartolini et al. (2012), (**d**) Variation of the CO₂ ppmV and δ¹³C in Greenland. Redrawn from McElwain et al. (2009). (**e**) Variation of the δ¹⁸O in SW Britain, numerical data by courtesy of Korte et al. (2009). Simplified from Guex et al. (2015)

The data allowing us to discuss the various hypotheses of recent extinction models and the timing concerning the End Triassic Extinction (ETE) and the T-J boundary is shown in Fig. 2.3 (for references see figure caption). This compilation synthesizes the timing of sea level changes, and $\delta^{13}C_{org}$, $\delta^{18}O$, and pCO_2 variations in relation with paleotemperatures, the age of the onset of the CAMP-related basaltic volcanism in the northeastern United States (Newark Supergroup) and Morocco (Argana Basin), and the ages of the two distinct End Triassic (ammonoids) and Earliest Jurassic (terrestrial plants) extinctions.

The chronology, established by ammonoids and U-Pb dating implies that the Newark supergroup basalts postdates the ETE and the disappearance of the latest Triassic ammonite Choristoceras crickmayi (Guex et al. 2004; Schoene et al. 2010). The delay between the recovery of the Jurassic ammonites and the extinction of the very last Triassic ammonoids lasted at least 200 kyr (probably more), based on sedimentary rates in Northern Peru and Nevada. The extinction of the last Triassic ammonoids in the uppermost Rhaetian is correlated with a strong negative excursion of $\delta^{13}C$ and a marine regression (Guex et al. 2004). The $\delta^{18}O$ record (Clemence et al. 2010; Korte et al. 2009) indicates a cooling episode, which could explain the regressive event recorded in the upper Rhaetian of Austria, England and Nevada. The initial regression is followed by a significant sea level rise potentially associated to large volcanic CO_2 emissions related to the CAMP basaltic volcanism (McElwain et al. 1999, 2009; Schaller et al. 2011; Bartolini et al. 2012). A major plant extinction is correlated with the greenhouse conditions that postdates the ETE by at least a few hundred thousand years. The plant extinction is recorded in Greenland (McElwain et al. 1999, 2009) and is associated with a second negative $\delta^{13}C$ recorded in the Hettangian Psiloceras planorbis beds (coeval with P. pacificum).

The recovery of the ammonites after the End Triassic extinction, calibrated with the geochronological data is illustrated in Fig. 2.4. This diagram shows a partial correlation between the $\delta^{13}C_{org}$ curve and the diversity fluctuations. This figure shows that the well-known first negative excursion of the organic carbon is correlated with the peak of the Rhaetian extinction.

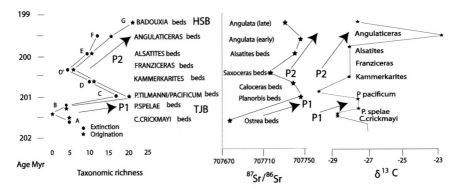

Fig. 2.4 Variations of the taxonomic richness of the ammonites and variations of $\delta^{13}C_{org}$ and $Sr^{87/86}$, from Guex et al. (2012a). P1 and P2: positive pulses of the carbon isotopes. TJB and HSB: Triassic Jurassic Boundary and Hettangian Sinemurian Boundary

The second negative excursion is restricted to the Psiloceras zone. The mid-Hettangian slow-down of the diversification is followed by an explosion of the diversity in the Upper Hettangian and by a new positive excursion of the organic carbon. However we note that the minimum of the ammonite diversity (D') occurs later than the minimum of the $\delta^{13}C_{org}$ curve which is located between the P. pacificum and Kammerkarites beds. The strontium data, based on Jones et al. (1994) measurements in Great Britain (see also Cohen and Coe 2007) are plotted as the mean-values of the original measurements per subzone. There is an apparent correlation between the taxonomic richness and the variation of the $^{87}Sr/^{86}Sr$ ratio. Such a correlation was already observed and discussed by Cardenas and Harries (2010) at a very large scale in the marine Phanerozoic genera. According to these authors, this correlation is basically controlled by the availability of marine nutrients.

2.2.3 The Pliensbachian -Toarcian Extinctions

A model similar to the one presented above (Fig. 2.3) can be proposed for the Pliensbachian Toarcian crisis which is known to be correlated with the onset of the Karoo-Ferrar large igneous province (Palfy and Smith 2000).

Recent high precision U-Pb dating on zircons of major sill intrusions in the Karoo basin can be directly correlated with the well-known Toarcian Oceanic Anoxic Event (OAE) and is concomitant with these sill intrusions into organic rich sediments of that basin (Guex et al. 2012b; Sell et al. 2014). In Fig. 2.5, we present a synthesis of major isotopic variations, the available geochronological data and major sea level variations. These data allow us to investigate whether and how the geochemical and biochronological data can be correlated with the magmatic activity of the Karoo-Ferrar LIP.

The end-Pliensbachian extinction, preceding the Toarcian AOE by a few hundred kyr (Dera et al. 2010), is marked by an important diversity drop (disappearance of 90 % of the ammonite taxa) associated with a generalized sedimentary gap linked to a marked regression event in NW-Europe and the Pacific area.

This regression was interpreted as being due to a major short lived glaciation (Guex et al. 2001, 2012b) coeval with the main extinction and preceding the main basalt eruptions. Our major arguments refer to an important emersion topography observed on seismic images of the North Sea (Marjanac and Steel 1997), to the evidence of polar ice storage (Price 1999) and to the deposition of thick conglomerates (Dunlap Formation in Nevada (USA) (Muller and Ferguson 1939) and Ururoa-Kawhia area, New Zealand (Hudson 2003)). The cooling model is supported by recent $\delta^{18}O$ data on belemnites (Gómez et al. 2008; Harazim et al. 2012) and by the discovery of glendonites in the upper part of the Pliensbachian (Suan et al. 2011). The origin of the major cooling is probably related to huge volcanogenic SO_2 degassing during the Late Pliensbachian preceding the major CO_2 emissions of the Early Toarcian (Guex et al. 2001).

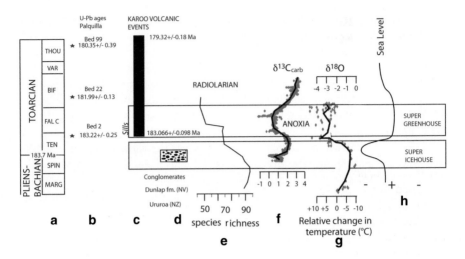

Fig. 2.5 A model for the Pliensbachian-Toarcian crisis. (**a**) Standard ammonite zonation of the Upper Lias. (**b**, **c**) Numerical ages of beds 2, 22 and 99 at Palquilla (Peru) and of the Karoo basalts. Redrawn from Guex et al. (2012b), Sell et al. (2014). (**d**) Conglomerates present at the Pliensbachian—Toarcian boundary in New Zealand and Nevada. From Guex et al. (2012b). (**e**) Species richness of Radiolarians in the Upper Pliensbachian and Early Toarcian. From Gorican et al. (2013). (**f**) Variation of the $\delta^{13}C$ around the Pliensbachian—Toarcian boundary. Redrawn from Hesselbo et al. (2007). (**g**) Variation of the $\delta^{18}O$ and paleotemperatures around the Pliensbachian—Toarcian boundary. Redrawn from Dera et al. (2010). (**h**) Variations of sea level around the Pliensbachian—Toarcian boundary. From Guex et al. (2001) and simplified from Guex et al (2015)

The regressive phase is followed, after a few hundred thousand years, by a world-wide transgression during the Early Toarcian, with the deposition of black shales associated with the Toarcian OAE (Jenkyns 1988). The Toarcian OAE itself is responsible for a second extinction affecting mainly benthic foraminifera populations (Bartolini et al. 1990) and brachiopods (García-Joral et al. 2011). Radiolarians were also affected (Fig. 4.5) but their extinction was apparently slightly delayed in comparison with benthos and probably coincided with a drastic fertility drop just after the OAE.

The succession of ice house conditions immediately followed by super green house conditions can be explained thanks to a petrological model elaborated by our colleagues Sebastien Pilet and Othmar Muntener to explain the SO_2 dominated vs. CO_2 dominated degassing couplet generating the successive cold and hot conditions. The model invokes a thermal erosion of the cratonic lithosphere inducing giant H_2S/SO_2 release from sulfur bearing basal continental crust before CO_2 becomes the dominant gas associated to the giant basalt emission (Guex et al. 2015b).

2.3   Graphic Representation of the Relationships Between Stress, Time and Evolutionary State

The catastrophe theory is a domain of the differential topology which was invented by René Thom (1972). It aims at building the simplest continuous dynamic model which can generate a morphology, given empirically, or a set of discontinuous phenomena.

Thom's theory concerns the phenomena where a gradual and relatively slow change produces a sudden jump of the state of the system. Such phenomena are called catastrophes. The graphical representation known under the name of "cusp catastrophe" is ideal to describe empirically the cases of the evolutionary jumps which arise during gradual changes in the environmental stress (Fig. 2.6). The surface illustrated in Fig. 2.3 represents the variable which characterizes the more or less advanced state of a taxonomic group which varies during the evolutionary time. This state is controlled by two parameters which, in our case, are the environmental stress and the time factor.

When these parameters vary, the curve of the state of the taxonomic group under study follows a trajectory which depends on time and on the intensity of the environmental stress. When the stress gradually reaches a certain threshold, the evolutionary state of the evolving system arrives at the border of the cusp and a jump occurs towards a previous state of more primitive aspect. In this book we will use such simple diagrams to describe the phenomena of retrograde evolution.

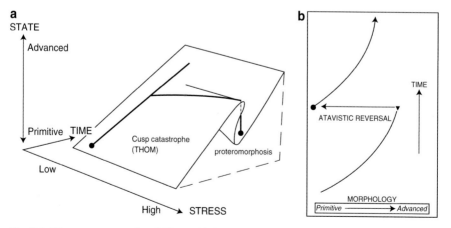

Fig. 2.6 The cusp catastrophe of Thom. (**a**) Stress and time are the parameters controlling the state of the evolving system (primitive–advanced). (**b**) Simplified graphic representation of (**a**)

Chapter 3
Retrograde Polymorphism in Some Planktonic Foraminifera

3.1 The *Ticinella-Thalmanninella* Lineage

A fundamental example of a Cretaceous planktonic foraminiferal anagenetic lineage starting from a very primitive form (evolute with simple rounded chambers) and giving rise to more complex forms with an involute and carinated shell was recently identified in the lineage *Ticinella-Thalmanninella* (Desmares et al. 2008) (Fig. 3.1). During the major Cenomanian oceanic anoxic event OAE2, the end forms of the lineage (group of *Thalmanninella greenhornensis*) gave rise to a very simplified atavistic group (*Thalmanninella multiloculata*, also called "*Anaticinella*"), which is a quasi homeomorph of its ticinellid ancestor. Here we present these discoveries (Desmares et al. 2008) in the light of what is known about the influence of high environmental stress on the development and variability of some other marine invertebrates.

3.2 Stratigraphy

The stratigraphic interval considered in the present section spans from the Late Albian to Late Cenomanian. There are records of two major environmental perturbations in this period: the Mid-Cenomanian Event and the Oceanic Anoxic Event called OAE2. These two anoxic events have markedly influenced the evolution of the *Ticinella-Thalmanninella* lineage as well as two diverging lineages represented by *Rotalipora montsalvensis-praemontsalvensis* and *montsalvensis-planoconvexa* (Gonzalez-Donoso et al. 2007). The stratigraphic distribution and outline of the phylogeny of these lineages are shown in Fig. 3.1 (from Desmares et al. 2008, modified; see also Caron 1985; Robaszynski and Caron 1995 for the stratigraphic details, and Guex et al. 2012b for discussion).

© Springer International Publishing Switzerland 2016
J. Guex, *Retrograde Evolution During Major Extinction Crises*, SpringerBriefs in Evolutionary Biology, DOI 10.1007/978-3-319-27917-6_3

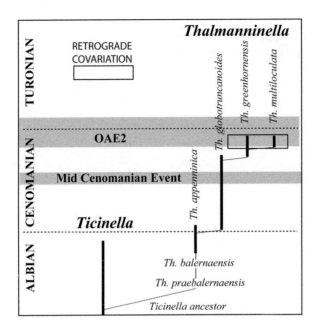

Fig. 3.1 Stratigraphic distribution and phylogeny of the Thalmaninellids discussed in the text (stratigraphic and phylogenetic data simplified from Desmares et al. 2008)

In Chap. 5 we will show that the anagenetic *Ticinella-Thalmanninella* lineage represents an example of geometrical transformation which is quite similar to what is known for many lineages of ammonites.

3.3 The *Thalmanninella*-"*Anaticinella*" Plexus

During the onset of the anoxic event OAE2, the intraspecific variability of *Thalmanninella greenhornensis* increases remarkably; this group gives rise to *Thalmanninella multiloculata* (Fig. 3.2), a "species" (morphospecies) with an indistinctly marked or absent keel, which is the last member of the *ticinellid-thalmanninellid* lineage (Desmares et al. 2008).

Th. greenhornensis is a complex trochospiral species that displays raised sutures on the spiral side, supplementary apertures on the umbilical side and a single keel. It further presents umbilical secondary apertures and non-inflated chambers on the umbilical side.

Globular morphotypes with supplementary apertures also occur in the same assemblages as the keeled ones. They were initially referred to *Anaticinella multiloculata* (Eicher 1972) which was first described in the North American Basin (Eicher 1972; Longoria 1973; Leckie 1985; Desmares et al. 2003).

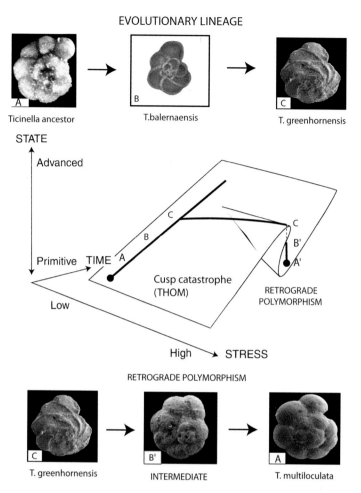

EVOLUTIONARY LINEAGE

Ticinella ancestor   T.balernaensis   T. greenhornensis

Fig. 3.2 Retrograde polymorphism affecting the *Thalmaninella* lineage during the upper Cenomanian anoxic event (data reinterpreted from Desmares et al. 2008; see Guex et al. 2012b)

In all Upper Cenomanian Western Interior Seaway outcrops, transitional specimens between the keeled forms of *Th. greenhornensis* and the globular morphotypes of *Th. multiloculata* were observed (Desmares et al. 2008). These morphotypes do not have a keel on all chambers of their final whorls. Where present, the keel is more or less pronounced, i.e. from indistinctly marked to thick and protruding. From *Th. greenhornensis* to *Th. multiloculata*, the chambers evolve progressively from crescentic to globular shape, the raised sutures become more depressed, and the periumbilical flanges disappear progressively on the umbilical side. The junction between the sutures and the periphery that is oblique in *Th. greenhornensis*, becomes perpendicular in *Th. multiloculata*. In lateral view, specimens of *Th. greenhornensis* that are compressed in form gradually change to a globular shape (Desmares et al. 2008).

Similar to the *Th. Greenhornensis-Th. multiloculata* plexus, numerous intermediate specimens were found between the keeled *Rotalipora cushmani* and the globular *Rotalipora planoconvexa* (details in Desmares et al. 2008). *R. cushmani* is also a
complex trochospiral morphotype that displays raised sutures on the spiral side,
supplementary apertures on the umbilical side and a single keel. *R. cushmani* has
sutural secondary apertures and equally biconvex chambers. While *R. cushmani*
displays a thick peripheral keel (Brönnimann and Brown 1955), *R. planoconvexa*
has globular chambers with an imperforate peripheral band (Longoria 1973). Raised
sutures on the spiral side of *R. cushmani* become depressed in the specimens of *R.
planoconvexa* and in transitional forms. Progressively, the single keel that is thick-
beaded and protruding in *R. cushmani*, is expressed less and less in forms transitional towards *R. planoconvexa*. On the umbilical side, the concentration of pustules,
typically forming a "Y" shape on each chamber of *R. cushmani*, disappears gradually on transitional forms and completely on *R. planoconvexa*. During the onset of
the Mid-Cenomanian event (Keller and Pardo 2004; Ando and Huber 2005), a similar transformation was also observed (Desmares et al. 2008). Under these environmental perturbations, the keeled species *R. montsalvensis* is associated with globular
morphotypes *R. praemontsalvensis*.

Chapter 4
Trend Reversals in Radiolaria During Extinction Periods

4.1 End Permian Event and the Entactinids

Since the Cambrian period when they first appeared, radiolarians have experienced several quasi extinctions but survive well until the present time. One of the most severe extinctions in their life history occurred at the end of the Permian and a careful review of the way they survived the Permian Triassic crisis was recently published (De Wever et al. 2006). It is demonstrated that following the Permian extinction some forms, such as the entactinarian *Parentactinia*, reduced the complexity of their skeleton by partial loss of the outer spherical shell that surrounds the spicule. Incidentally, many earliest Triassic spicular forms belonging to the entactinids are the result of a loss of the outer skeleton (De Wever et al. 2006). The evolutionary process leading to simple and primitive looking end-forms under the influence of high environmental stress in radiolarians is very similar to what is observed in the Silicoflagellids and Foraminifera. Such simplifications often correspond to the reappearance of ancestral geometries in the *Ticinella* lineage discussed above. Three examples illustrating such phenomena are given below.

4.2 Retrograde Evolution of the *Albaillella* Lineage During the Permian-Triassic Crisis

The first detailed evolutionary study of the genus *Albaillella* in the Upper Permian of Japan demonstrated that the *Albaillella* lineage starts with a small form having a conical shell with six narrow transverse bands and a ventral wing called *A.* sp. *G* (Kuwahara 1999 and Fig. 4.1). The apical part of the shell of this ancestral form is almost straight with a ventral wing protruding from near the last transverse band. It is followed by *Albaillella yamakitai* Kuwahara, which has a conical shell with five transverse bands, a dorsal bulge and one ventral wing. Next in the sequence is

© Springer International Publishing Switzerland 2016
J. Guex, *Retrograde Evolution During Major Extinction Crises*, SpringerBriefs
in Evolutionary Biology, DOI 10.1007/978-3-319-27917-6_4

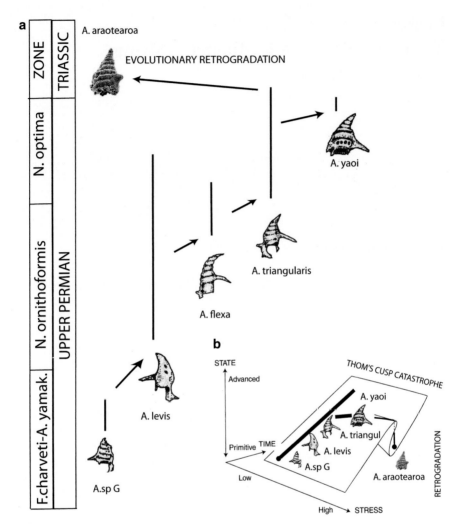

Fig. 4.1 (**a**) Anagenetic evolution of *Albaillella* during the *Upper* Permian showing increasing convexity of the conical shell and reversal of the trend above the Permian-Triassic boundary. (**b**) Diagrammatic representation of the evolutionary catastrophe observed in the Albaillellids at the Permian Triassic transition (from Guex et al. 2014)

Albaillella sp. *A.* with a conical shell and one ventral wing with dimple-like holes. The shell is small, conical, flattened with a ventral wing and a strongly pentagonal form. The upper part of the shell starts to be slightly curved on the ventral side. *A.* sp. A is followed by *Albaillella protolevis* Kuwahara and *A. levis* Sashida and Tonishi, characterized by a conical, smooth and inflated shell with one ventral wing.

   Albaillella lauta Kuwahara shows a ventrally curved upper part and shell height is greater than in *A. levis*. The next one, *A. flexa* Kuwahara, has a ventro-apical shell that is remarkably curved. It is followed by *A. angusta*, which is slightly less curved and by *A. excelsa*, which is smooth and elongated. The last three representatives of the lineage, *A. triangularis*, *A.* sp. D and *A. yaoi* show extreme curvature of the apical part of the cone.

To summarize, the phyletic sequence is gradual with most intermediate forms represented and with the derived forms replacing progressively the ancestral forms. In other words there is no place for a persistent opportunistic form surviving during the development of the lineage. Of note, Kuwahara's uppermost Permian section is cut by a fault. Nevertheless, we have a perfect record of the geometrical transformations of *Albaillella* during that critical evolutionary interval. Following the greatest extinction in the history of the Earth, the oldest Lower Triassic record of *Albaillella*, is provided by Takemura and Aono (2007) with the discovery of *Albaillella aotearoa* Takemura at Arrow Rocks (Oruatemanu), New Zealand. This Triassic species has a small, somewhat flattened conical shell with a ventral wing. The upper part of the shell is almost straight in outline and is clearly a homeomorph of *Albaillella* sp. G and *A. yamakitai* described by Kuwahara at the beginning of the Upper Permian lineage, providing a nice example of morphological retrogradation (=proteromorphosis).

4.3   The Evolution of the Saturnalids During the KT Crisis

The fundamental studies on the evolution of the oertlispongids and their direct derivatives, the saturnalids, provide an outstanding illustration of the long-term complexification of some radiolarian skeletons (Dumitrica, 1982, 1985). In the Early Paleocene, which immediately followed the End Cretaceous Extinction, this group was marked by the appearance of a very simplified radiolarian group, the Axopruninae, a saturnalid that lost its equatorial ring and retained only two polar spines. The Triassic oertlispongids are represented by a simple spongy spherical shell with two spines. The upper spine of these radiolarians was transformed into an arch with increasingly complex geometry, suggesting at the same time an affinity of this group with the saturnalids (Dumitrica, 1982). Subsequently it was established that Lower Jurassic saturnalid evolution is characterized by the transformation of the spongy inner shell into a sequence of two concentric latticed medullary shells, surrounded by an external shell. Further extrapolation of these initial discoveries showed that saturnalids derived from the oertlispongids by a similar complexification of the lower spine, followed by a fusion of the two arcs, leading to the genesis of typical saturnalids with an equatorial ring (Kozur and Mostler, 1983, 1990). Similarly several Cenozoic forms with simple subspheric cortical shells with two axial spines (*Axoprunum*, *Xyphosphaera*) derived from the usual saturnalids by loss of the equatorial ring, a characteristic of this large family (Dumitrica, 1985). This major geometrical simplification occurred suddenly within the Lower Paleocene and it clearly results from the major crisis leading to the Cretaceous-Tertiary extinction. The loss of the equatorial ring is comparable with the loss of the basal ring of the silicoflagellids at times of artificial or natural ecological stress (Fig. 4.2).

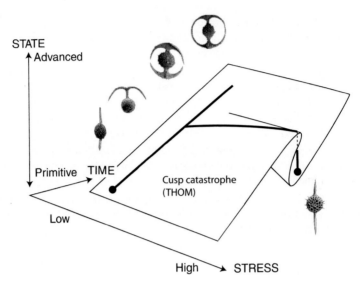

Fig. 4.2 Idealized view of the evolution of the saturnalids and loss of the equatorial ring at the Cretaceous-Cenozoic boundary. Not to scale (constructed after a diagram of Guex 1993)

4.4   Untwisting of Spines Around the Triassic–Jurassic Boundary Crisis

Drastic retrogradation in spine geometry occurs during the Triassic–Jurassic Boundary crisis. *Tipperella* is a late Norian to Hettangian genus characterized by having a simple subspherical test with four spines in the tetrahedral position. The cortical shell wall is thick and variable in appearance, composed of either small polygonal pore frames or spongy meshwork. The interior cavity is filled with spongy meshwork that is frequently dissolved. Spines are triradiate and strongly twisted through most of the Rhaetian stage (see Fig. 4.3), but towards the end of it they begin to untwist. This contrasts remarkably with basal Hettangian forms whose spines are straight and circular in cross section, similar to many Permian radiolarians.

A similar case of untwisting can also be seen in *Betraccium* during the TJ crisis (see Guex et al. 2012). This middle Norian pantanelliid has a subspherical cortical shell with coarse polygonal meshwork and three radially arranged primary bladed spines in the same plane. *Betraccium smithi* Pessagno and other Norian species such as *B. deweveri* Pessagno and Blome, *B. maclearni* Pessagno and Blome and *B. yak-ounense* Pessagno and Blome have twisted spines. All species with strongly twisted spines disappear around the Norian–Rhaetian boundary, and all subsequent species have straight three-bladed spines (Carter, 1993).

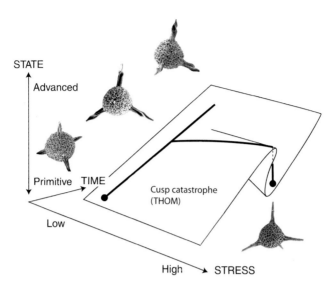

Fig. 4.3 Loss of the twisting of the spines in *Tipperella* at the Triassic–Jurassic boundary. Diagram, not to scale (from Guex et al. 2012a)

4.5 The Evolution of *Eucyrtidiellum* During the Pliensbachian Toarcian Stages

In this section we will study the evolution of the genus Eucyrtidiellum during the Lower Jurassic and its reaction during the Pliensbachian Toarcian major crisis. To give a clear picture of the rate of faunal turnover among radiolarian during that period, we expressed it as the number of species with FAD (first appearance datum) against the number of species with LAD (last appearance datum) in a given biochronological unit, called Unitary Association (see Guex et al. 2015a, b). A special tool of the UAgraph program (Hammer et al. 2015) called "Cumulated FADs/LADs" was used to construct Fig. 4.4 and we used the cumulative number of FADs plotted against the cumulative number of LADs. Each Unitary Association is represented by a point on the curve. Gentle slopes of the curve indicate high diversification rates (great number of FADs vs. low number of LADs) and steep slopes indicate high extinction rates (low number of FADs vs. great number of LADs). Take notice of the problem that the beginning and the end of such a curve are biased by the fact that the base and the top of the original range chart record truncated ranges: all taxa in the lowest biochronological unit may range downward and all taxa in the highest unit may range upward. These parts of the curve must obviously be ignored. The advantage of this method is that it analyses the relationship between FADs and LADs and not only their absolute numbers. This means that the analysis is not biased by exceptional preservation, which is a common phenomenon in the radiolarian fossil record.

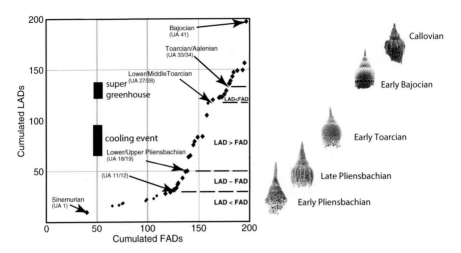

Fig. 4.4 Cumulated FADs vs. cumulated LADs of the radiolarian during the Sinemurian-Bajocian interval and the evolution of Eucyrtidiellum during the Pliensbachian-Callovian interval. Quantitative stratigraphic data from Carter et al. (2010), Gorican et al. (2006) and Gorican et al. (2013)

Eucyrtidiellum is a very common Jurassic nassellarian looking like a tiara. In the Pliensbachian *E. nagaiae* with an abdomen typically ornamented with strong vertical costae, evolves from *E. gunense* (Gorican et al. 2006). During the topmost Pliensbachian–Early Toarcian major crisis, these forms give rise to smooth derivatives without the abdominal costae (*E. disparile*), that evolve to smooth poreless forms (*E. unumaense*) in the early Middle Jurassic. Abdominal costae gradually reappear in the Bathonian (*E. dentatum—E. pustulatum—E. semifactum, E. ptyctum* and later *E. pyramis*). The pores on the abdomen progressively close in this lineage.

The drastic simplification of the ornamentation and retrograde evolution of Eucyrtidiellum during the Pliensbachian-Toarcian crisis is clearly related to the major environmental perturbation which occurred during that period (Gorican et al. 2006). That crisis is also well known to be correlated with the onset of the Karoo-Ferrar large igneous province discussed in Sect. 2.2.

Chapter 5
Evolution of Some Cephalopods During Major Extinctions

5.1 Repetition of Lineages by Retrograde Evolutionary Jumps in Ammonites

Iterative evolution is the most striking property of the ammonoid distribution during their long life. The very large scale phylogeny of the three major distinct groups occurring during the Paleozoic and the Mesozoic periods is shown in Fig. 5.1 which demonstrates that major evolutionary jumps in ammonoids occur during severe extinction events. These are characterized by the rapid appearance of simple, primitive-looking forms which are similar to remote ancestors of their more complex immediate progenitors. Such forms are often atavistic and homeomorphic species generated during such sublethal stress events and they can be separated by several millions of years from their initial ancestor. A striking example of such an iteration is given in Fig. 5.2 presenting the very simple morphology of the smooth evolute (or *Xenodiscus* group in a larger sense), appearing at the Permian Triassic boundary and giving rise to all the ammonoids known as Ceratitina. The group is affected by a quasi extinction at the Triassic Jurassic boundary and its recovery starts from the equally simple smooth evolute *Psiloceras*, which are, in turn, quasi homeomorph of *Ophiceras*. These two primitive forms developed phyletic lineages with many common features (Fig. 5.2), even though *Psiloceras* is derived from *Ophiceras* through a long and complex series of successive and distinct lineages. The detail of the relation between Phylloceratina and Psiloceras is given in Fig. 5.3.

© Springer International Publishing Switzerland 2016
J. Guex, *Retrograde Evolution During Major Extinction Crises*, SpringerBriefs
in Evolutionary Biology, DOI 10.1007/978-3-319-27917-6_5

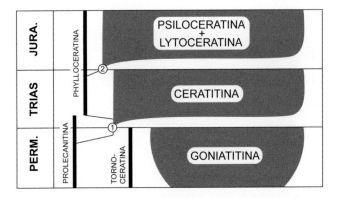

Fig. 5.1 Large scale evolution of the different suborders of ammonoids. (1) Ophiceras, (2) Psiloceras (see text and Sect. 2.2). From Guex (2006)

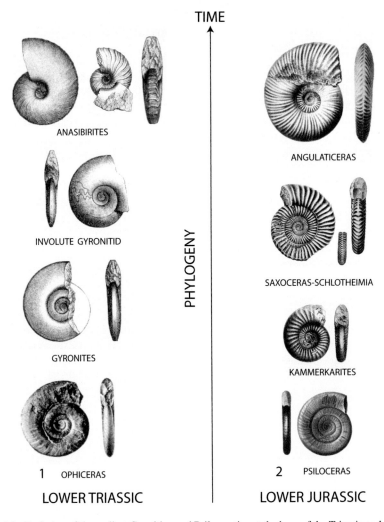

Fig. 5.2 Evolution of the earliest Ceratitina and Psiloceratina at the base of the Triassic and of the Jurassic. From Guex (2006)

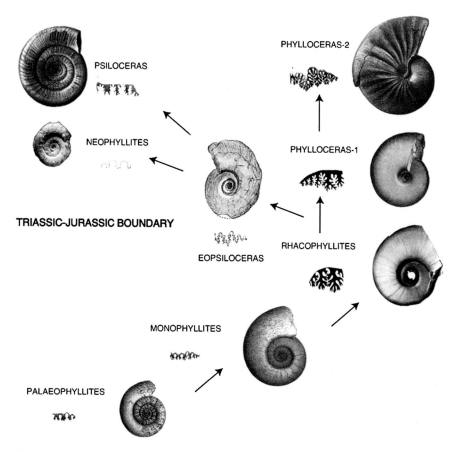

Fig. 5.3 Details of the relations between the Phylloceratida and Psiloceras (from Guex 2006). See Sect. 2.2 and Fig. 5.1

5.2 Evolutionary Jumps of the Ammonites During the Middle-Late Toarcian Crisis

Here we return to two examples which we consider particularly illustrative (loc. cit.). The first concerns the evolution of the Middle Toarcian *Paronychoceras* (the microconchs of *Phymatoceras*, see Guex 2000) and of their Late Toarcian offshoots, *Onychoceras* (the microconchs of *Hammatoceras*) (Fig. 5.4). The oldest known *Paronychoceras, P. pseudoplanum*, is a tiny relatively evolute smooth form with no keel or sign of ribbing. It is followed by *P. costatum*, which has no keel but faint crescentic lateral ribs. Its direct offshoot is represented by *Pseudobrodieia lehmanni*, itself followed by *Brodieia* sp. n. ind. which is ribbed, keeled and has lateral lappets. In the Early Late Toarcian (uppermost Variabilise zone-Thouarsense zone) the first microconch lineage gives rise to *Onychoceras planum*, followed by

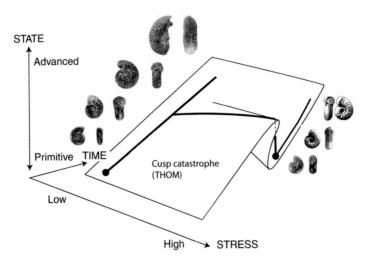

Fig. 5.4 Evolution of the *Paronychoceras → Pseudobrodieia → Brodieia* lineage and catastrophic repetition of the early part of the lineage at the Middle-Late Toarcian boundary (from Guex 2001)

O. tenue (Early Insigne Zone) and *O. differens* (Late Insigne Zone). It is of importance that no intermediate forms linking the two successive lineages are known (Fig. 5.4). This possibly reflected the poor preservation of this group during the Middle-Late Toarcian crisis.

A second lineage of interest is observed during the same time which shows a similar case of iterative evolution: the *Hildaites → Phymatoceras → Haugia → Denckmannia* (Middle Toarcian Phymatoceratinae) which gives rise to the plexus *Podagrosites → Pseudogrammoceras → Phlyseogrammoceras → Huddlestonia* lineage (Upper Toarcian Grammoceratidae). The evolution of the Phymatoceratinae and one of their derivatives, the Grammoceratidae, is illustrated in Fig. 5.5. This lineage is rooted in evolute forms of the Lower Toarcian which belong to the genus *Hildaites*, and which evolved into the involute group of *Haugia* via *Phymatoceras*. During the Late-Middle Toarcian, the variability of *Haugia* increased and this group gave rise to a relatively evolute form called *Denckmannia* which, in turn, gave rise to a very evolute and simply ribbed ammonite, *Podagrosites*. This new group is a perfect homeomorph of the ancestral *Hildaites*. Starting from the *Podagrosites* pole, we observe a continuous morphological transition towards the different species of *Pseudogrammoceras*, eventually giving rise to the involute and fasciculate *Phlyseogrammoceras* and to the smooth oxycone *Hudlestonia*. This example demonstrates that the transition from involute forms (which are the outcome of a major evolutionary trend) towards a primitive looking evolute form is achieved through an increase of the variability of the ancestral advanced group during an episode of stress. It should be noted that such a transformation is global (i.e. it affects the whole ontogeny) and does not result from a simple heterochrony such as neoteny or progenesis.

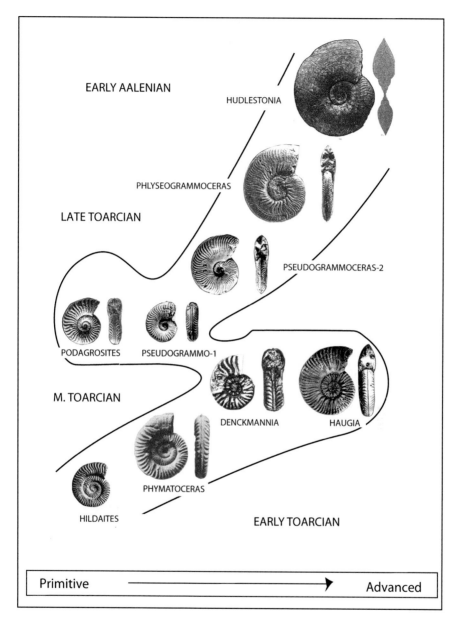

Fig. 5.5 Anagenetic evolution of the *Hildaites → Phymatoceras → Haugia* lineage and catastrophic appearance of the sub-homeomorphic *Podagrosites → Pseudogrammoceras → Phylseogra mmoceras* lineage (from Guex 2001)

5.3   Retrograde Covariation During Extinction Episodes

5.3.1   Buckman's First Law of Covariation

The *Buckman's First Law of covariation* named so by Westermann (1966) following the first observations by Buckman (1887) in *Sonninia* and *Amaltheus* was
originally described as follows: "Roughly speaking, inclusion and compression of
the whorls correlate with the amount of ornament—the most ornate species being
the more evolute (i.e. loosely coiled) and having almost circular whorls…"
(Buckman 1887).

Some Pliensbachian ammonites belonging to the genus *Amaltheus* provide probably the best example of covariation. In this group, more specifically in the *A. gibbosus* group, all transitions between evolute and strongly spinose forms and
quasi-smooth involute forms at the other extreme are observed (Figs. 5.6 and 5.11b),
excluding the possibility to explain this compelling variability as a special case of
scale and proportionality (Hammer and Bucher 2005).

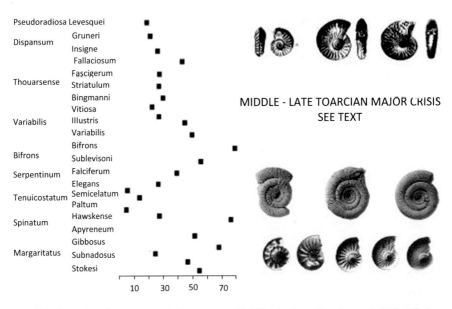

Fig. 5.6 Examples of covariation during the unstable Pliensbachian-Toarcian period: Correlation
between the development of extreme retrograde polymorphism and main extinction events (see
variations of the species richness). Margaritatus zone: *Amaltheus gibbosus*. Tenuicostatum Zone:
Dactylioceras clevelandicum (courtesy of Mike Howarth). Dispansum Zone: *Osperleioceras
reynesi*

5.3.2   *The Case of* Osperleioceras reynesi

As already seen above, the crisis periods are not favourable to fossilization. During such perturbated intervals, the fossils are absent, rare or badly preserved. However, during the Middle Upper Toarcian crisis, the genus *Osperleioceras* is very well recorded in the Causses Basin (Southern France) and we can follow precisely the genesis of the loosely coiled and atavistic *Osperleioceras reynesi* from its tightly coiled ancestor *O. bicarinatum* which itself derived from the common ancestor Harpoceras.

The details of that very important morphological genesis of a primitive looking form are illustrated in Fig. 5.7. Note that the last step of that evolution is marked by a clear phenomenon of covariation, discussed below.

5.3.3   *A Morphogenetic Explanation of Buckman's Covariation*

Covariation depends on the internal shell geometry, namely the lateral and ventral curvature of the shell which controls the amount of morphogens present in the more or less curved mantle, the most salient ornamentation being present where the whorls are most curved, shells with slight angular bulges often being spinose or carinate and flat ones being almost smooth (Guex 1999, p 42). The empirical conclusion was that the covariation phenomenon could be explained within the framework of Gierer-Meinhardt's reaction diffusion models. To prove that conclusion, André Koch simulated the distribution of "morphogens" (in the physical sense) in a quadrangular body chamber and demonstrated that morphogens maxima are located, as expected, in the part of the mantle located in the angular parts of the shell.

For this, Andre Koch (in Guex et al. 2003) calculated a numerical solution of the Gierer-Meinhardt equations for a cross section through an ammonite shell, orthogonal to the growth axis. The solution of the standard reaction-diffusion equations of Gierer-Meinhardt in a bidimensional domain is as follows:

$$\partial_t a = D_a \Delta a + \rho_a a^2 / h - \mu_a a + \sigma_a$$

$$\partial_t h = D_h \mathrm{D}h + \rho_h a^2 - \mu_h h$$

with $\Delta \equiv \partial^2 / \partial_x^2 + \partial^2 / \partial_y^2$; $a(x,t)$ and h(x,t) corresponding to the activator and inhibitor morphogens, respectively. In the numerical simulation, the constants have the following values:

$$D_a = 0.012 \quad \rho_a = 1.0\, \mu_a = 0.01 \quad \sigma_a = 0.002$$

$$D_h = 0.4\rho_h = 1.0\mu_h = 0.01$$

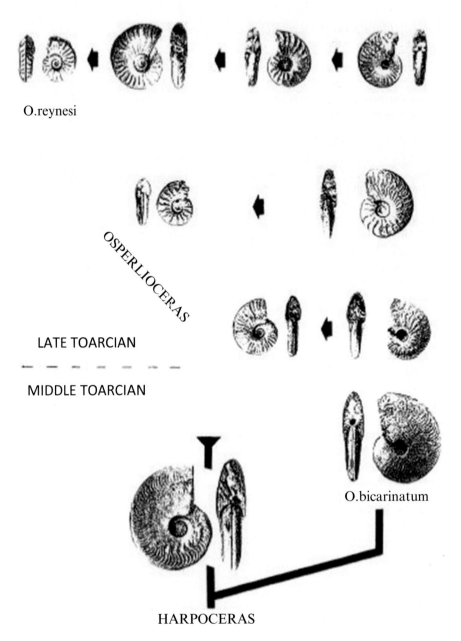

O.reynesi

OSPERLIOCERAS

LATE TOARCIAN

MIDDLE TOARCIAN

O.bicarinatum

HARPOCERAS

Fig. 5.7 Evolution of *Osperleioceras* through the Middle-Late and Late Toarcian crises, showing the development of a retrograde polymorphism in the Reynesi subzone. From Guex (1992)

The units of distance, time and concentration are arbitrary! The domain of computation corresponds to the union of the two following areas:

1. Area delimited by two arcs of circles {(−33.5120,−12.1072), 50.00} and {(−24.5332, −11.5931), 40.00}, and contained in the region $x \geq 0$ and $y \geq -0.052$.
2. Area delimited by two arcs of circles {(33.5120, −12.1072), 50.00} and {(24.5332, −11.5931), 40.00}, and contained in the region $x \leq 0$ and $y \geq -0.052$.

The boundaries of the domains are supposed to be impervious for the inhibitor. The outer boundary (arc of circle of radius 50.00) is unaffected by the activator whereas the other boundaries are susceptible to this factor. The choice of these boundary conditions is motivated by the following arguments: The activator is supposed to diffuse freely outside the mantle's cells into the environment (intercellular medium and sea water).

The reaction-diffusion equations are solved numerically on a hexagonal mesh containing 1500 nodes corresponding to a hexagon radius of 0.23 units. The concentrations $a(x,t)$ and $h(x,t)$ are determined at each node of the mesh. The initial values of the concentrations at $t=0$ correspond approximatively to the values taken from the (unstable!) homogenous stationary solution. We add small random deviations $\varepsilon(x,0)$ to the concentrations of the activator to allow the system to leave the initially homogenous state. The initial values are thus given by:

$$a(x,0) = 1.0\left[1 + \varepsilon(x,0)\right] \quad \text{with} \quad -0.05 < e(x,0) < +0.05$$

$$h(x,0) = 100.0.$$

The stationary inhomogeneous solution is found using a standard iterative procedure (details in Guex et al. 2003). Similar conclusions were obtained by Newell et al. (2008) in their general study of phyllotaxy: "… buckling leads to a template for primordia, it is growth that leads to the visible primordial bumps and phylla. This growth is postulated to be a biochemical response, perhaps through chemical agents such as auxin, to the local stress or curvature inhomogeneities of the buckled surface …". In our carbonate shelly invertebrates, the morphogens have obviously nothing to do with auxin but could simply be Ca^{2+} ions.

The output of Koch's calculation is given in Fig. 5.8, showing the distribution of the activator and inhibitor in a bended shell: very low concentration in the smooth part and very high concentration in the curved part.

More recently, an alternative model leading to the same kind of conclusions has been proposed by Mercker et al. 2013 where the authors write that "biomechanical forces may replace the elusive long-range inhibitor and lead to formation of stable spatially heterogeneous structures without existence of chemical prepatterns. We propose new experimental approaches to decisively test our central hypothesis that *tissue curvature and morphogen expression are coupled in a positive feedback loop*".

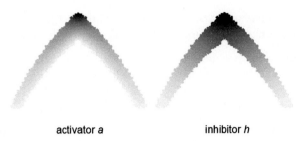

activator *a*   inhibitor *h*

Fig. 5.8 Output of Koch's program showing the relative concentration (*white to dark grey*) of the activator and inhibitor in the bended shell (see text). From Guex et al. (2003)

It might be useful to note that a multitude of mathematical models can simulate the pattern formation in shelly organisms. To illustrate this we can mention the fact that phyllotaxy, a domain studied by botanists, has been formally described by models developed in three totally different fields of mathematics: pure geometry by Van Iterson (1907), pure physics by Douady and Couder (1992) and reaction-diffusion by Meinhardt et al. (1998). In other words it is clear that our model of spine formation in ammonites (Fig. 5.8) is obviously not the only possible, even if it has nothing to do with a proportionality problem.

5.3.4   Covariation and Environmental Stress

One notable characteristic of covariation in Buckman's sense is that it usually generates extreme morphotypes of atavistic habitus with simple and evolute morphologies and it seems to occur exclusively during episodes of environmental stress.

To demonstrate this we shall stay in the upper part of the Lower Jurassic (Late Pliensbachian-Toarcian), a period of high ecological instability that provides evidence of some of the best documented extinction events in ammonoid history.

The variations in ammonite biodiversity, expressed as species richness, from the Pliensbachian-Toarcian boundary up to the lower part of the Late Toarcian (Dera et al. 2010) are correlated with the standard ammonite biochronological scale (Fig. 5.6).

The first extinction event is placed at the transition between the Gibbosus and Spinatum zones in the Late Pliensbachian (Meister 1988, 1989; Meister and Stampfli 2000; see also Dera et al. 2010). The evolute spinose zonal index *Amaltheus gibbosus* is an atavistic form generated from involute *A. margaritatus* which strongly resembles the ancestral morphology of the Sinemurian *Eoderoceras* (Guex et al. 2003, and see Fig. 5.11).

Across the Pliensbachian-Toarcian boundary (Fig. 5.6) the Dactylioceratidae reacted similar to other groups by a marked increase in variability with extreme forms leading from the depressed and strongly spinose *Dactylioceras crosbeyi* group to serpenticone forms with simple ornamentation of the *Eodactylites* morphogroup, another typical case of covariation.

The next crisis, known as the Early Toarcian Oceanic Anoxic Event, is characterized by the occurrence of the same kind of variability increase within the *Dactylioceras semicelatum* and *D. clevelandicum* group which develops spinose forms with cadicone internal whorls (Howarth 1973). This episode is also characterized by the appearance of evolute Hildoceratids known as *Hildaites*, derived from the relatively involute Protogrammoceratids. In this latter case the complete transition between the two groups is not well recorded.

During the Middle-Late Toarcian transition, an exceedingly important faunal turnover is observed with the disappearance of the Hildoceratids, Dactylioceratids and Phymatoceratids which are replaced in the ammonite populations by the Hammatoceratids and Grammoceratids. This crisis, like the others, is marked by a strong increase in polymorphism of the ammonites and is also marked by the appearance of atavistic/evolute primitive looking ammonites such as *Podagrosites* generated by *Pseudogrammoceras* (see above), which are similar to the ancestral lower Pliensbachian *Fuciniceras* (Guex 1992, 2006).

The same kind of biotic situation occurs once more during the Late Insigne–Early Levesquei crises (Guex 1975; Dera et al. 2010) where the disappearance of abundant *Osperleioceras*, most typical Hammatoceratids, *Alocolytoceras, Buckmanites* and *Oxyparoniceras* is observed. In the topmost abundantly fossiliferous beds of the Late Insigne Zone (Dispansum subzone) a huge polymorphism is observed again in the *Osperleioceras reynesi* group with typical *Osperleioceras* like *O. alterans-wunstorfi* (Guex 1975, Pl. 8, Fig. 3) giving rise to simplified *O. reynesi* (Guex 1975, Pl. 8, Fig. 2), a form characterized by an evolute coiling and simple, strong and almost straight lateral ribs (Guex 1992, Fig. 4; Morard and Guex 2003, Fig. 1). The same is detected in *Hammatoceras* of the *bonarellii* group (Guex 1975, Pl. 9, Fig. 12) which gives rise to serpenticone *Catulloceras* (Guex 1975, Pl.2, Fig. 4), the direct ancestor of the *Dumortieria* and *Pleydellia* leading to all the Haplocerataceae of the Middle and Upper Jurassic (Guex 1992, Fig. 5). Other examples of atavisms occurring during high environmental stress are discussed and illustrated in Guex (2006).

5.3.5 Ecophenotypes and Stress in Benthic Foraminifera

Benthic foraminifera are well known to be extremely sensitive to environmental stress (Camacho et al. 2015; Barras et al. 2010) and several classical studies have shown that the variability of these organisms was polarized in function of such perturbations. That polarization is mainly characterized by a differential colonization of the ecological niches where the geometrically simplest and archaic forms are more frequent in unstable environments. In a fundamental contribution, Grünig (1984) has demonstrated the influence of the depth over the different phenotypes of Spiroplectammina and Uvigerina in which the forms with spinose ornamentation predominate in deep environments when relatively smooth forms colonized the shallow waters (Fig. 5.9).

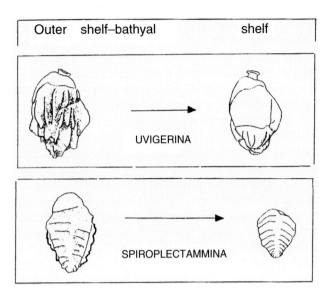

Fig. 5.9 Influence of depth over the different phenotypes of *Spiroplectammina* and *Uvigerina* (From Grünig 1984)

Fig. 5.10 Morphocline showing the extreme forms going from the disorganized glomospiroid type to the typical planispiral *Hemigordius* type in function of environmental stress. Modified from Gargouri and Vachard (1988) in Guex (1992)

On another hand an extreme variability in the Permian porcelaneous foraminifera such as *Hemigordius* (Fig. 5.10) was demonstrated by Gargouri and Vachard (1988) and Mehl and Noe (1990). In this group we observe all the transitions between glomospiroid forms and regular evolute panispiral morphotypes (Fig. 5.10). The glomospiroid morphotypes predominate in the hypersaline waters where nothing else can survive.

5.3.6   Comparison Between the Covariation Observed in Metazoans and Foraminifera

The evolutionary trend observed in the *Ticinella → Thalmaninella* lineage discussed in Sect. 3.3 (Fig. 3.2) illustrates the frequent evolutionary trend occurring in coiled shells: in the *T. greenhornensis–T. multiloculata* plexus a strong keel is developed at

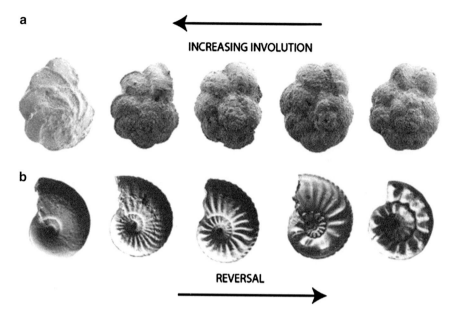

Fig. 5.11 (**a**) Evolutionary lineage going from the simple and evolute ancestral planktonic fora-minifera *Ticinella* towards the involute and carinated *Thalmaninella*. (**b**) Highly evolute "*Eoderoceras*-looking" (=ancestral) *Amaltheus* showing a similar trend from evolute towards invo-lute coiling

the maximum lateral curvature of the chamber, demonstrating the relation between morphogen concentration and strong curvature of the membrane. During episodes of environmental stress such as anoxic events, a retrograde polymorphism is observed in these protists, which is perfectly similar to the case of *Amaltheus* described in the preceding section, as illustrated in Fig. 5.11b.

The fact that Buckman's First Law of Covariation applies to both unicellulars and metazoans proves that similar biochemical signals are at work in both types of organisms. This note will be useful for our concluding remarks (see Chap. 9).

5.3.7 Heteromorph Ammonites and Uncoiling of Nautilus Above the Permian Triassic Boundary

The contemporary *Nautilus pompilius* is often regarded as one of the most typical example of a living fossil. Indeed, many very old (up to 300 Ma) nautiloids (s.l.) have a shell geometry which is very similar to that of the modern forms. It is there-fore surprising to see that this ultraconservative group has been able to behave like ammonoids during major crises (see Fig. 5.12).

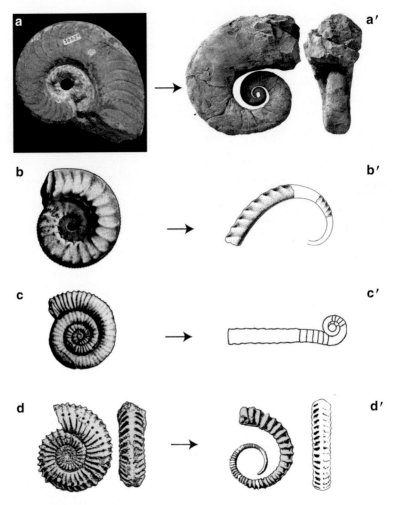

Fig. 5.12 (**a**) *Domatoceras* (Permian), (**a′**) *Gyronautilus* (Early Triassic. From Shigeta 2009), (**b**) *Hybonoticeras* (Tithonian), (**b′**) *Protancyloceras* (Tithonian), (**c**) *Cycloceltites* (U.Norian) (**c′**) *Rhabdoceras* (Rhaetian), (**d**) *Strenoceras* (Bajocian), (**d′**) *Spiroceras* (Bajocian) (**b–d** from Guex 2001)

The unique and fascinating example given here was found by Shigeta (2009) who described an uncoiled nautiloid at the base of the Triassic, deriving from the Pennsylvanian–Permian normally coiled *Domatoceras* (Fig. 5.12A). That extraordinary transition occurred during the catastrophic Permian Triassic transition, which, again, is known to be one of the worst crisis ever occurring on the Earth with 95 % of the species being extinct.

The most spectacular reversals of the trends towards increasing involution are observed among ammonites called "heteromorphs" (uncoiled or with helicoidal coiling) which appear abruptly at various periods of the Triassic and of the Jurassic and which proliferate during the whole Cretaceous.

The oldest uncoiled Mesozoic group belongs to the genus *Rhabdoceras* and derives from the normally coiled *Sympolycyclus/Cycloceltites* group at the base of the Cordilleranus zone, just at the end of a major regressive event. The straight conical *Rhabdoceras* generates semiuncoiled ammonites such as *Choristoceras* (Spath 1933), *Peripleurites* and the helicospiral *Cochloceras*. The Late Triassic heteromorph *Choristoceras* proliferated during the whole Rhaetian when most other typical Triassic ammonoids disappeared. Similar occurrences of uncoiled ammonites are observed during the Middle and Late Jurassic with the appearances of *Spiroceras, Parapatoceras*, discussed below with more detail, and *Protancyloceras*, deriving from normally coiled ammonites.

Magaritz (1989) was one of the first to understand the relationship between some carbon negative shifts ($\delta^{13}C$) and major extinctions in marine faunas. Indeed, it is obvious that collapses of biological productivity occurring during such periods imply massive release of light carbon which, in turn, appears in the stratigraphic record. Taken alone, negative shifts of the $\delta^{13}C$ are not sufficient to deduce the existence of environmental stress but they certainly can be taken as a stress index when they are concomitant with marine regressions and relatively high extinction rates.

In the case of the appearance of *Spiroceras* and *Parapatoceras*, we observe that they are precisely generated during two successive regressive events associated with negative shifts of the $\delta^{13}C$ and clear drops in ammonite diversity (Fig. 5.13). These events could be related to the major changes in the Pacific plates during Middle Jurassic and to the volcanism generated by these events (Bartolini and Larson 2001). The first one appears in the Blagdeni-Banski subzones and derives from lineage *Su bcollina → Parastrenoceras → Strenoceras* which are rooted within more involute Stephanoceratidae. The second one first occurs in the Orbis zone and derives from the lineage *Cadomites* (relatively involute), *Hemigarantiana*, *Epistrenoceras* (evolute).

Once more this uncoiling is a global process: uncoiled ammonites do not at all resemble juveniles of their ancestors and, therefore, they cannot be considered as "paedomorphs". It is also worth noting that such peculiar groups are perfectly well adapted to the unstable rhaetian and cretaceous environments. For example, one of the only survivor of the End Triassic Extinction (ETE) is *Choristoceras* which survived until the early Hettangian Planorbis zone.

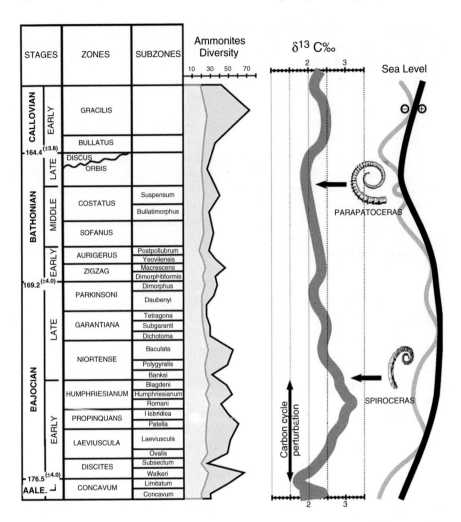

Fig. 5.13 Appearance of the Middle Jurassic *Spiroceras* and *Parapatoceras* during regressive phases and negative anomalies of the δ¹³C. From O'Dogherty et al. (2006)

Chapter 6
Retrograde Evolution of Corals at the End Triassic Crisis

6.1 Introduction

Rugose corals, characterized by their strong bilateral symmetry and complex skeleton, were amongst the most successful animals in the Paleozoic shallow waters, represented by several tenths of species. During the Permian-Triassic crisis, they have been obliterated and the Triassic recovery of reef building by scleractinian corals with hexagonal radiate symmetry took several millions of years. In 2003, Stanley proposed that these corals lost their shelly skeleton and had a planktonic way of life. Later the same hypothesis was introduced to explain that the change of symmetry could have been acquired because of this new mode of planktonic life (Guex 2006).

6.2 Coral Reefs During the End Triassic Extinction

During the End Triassic Extinction (ETE) the scleractinian corals, which were important reef builders during the Upper Triassic, underwent a marked decline that was followed by a "reef gap" during the Hettangian and early Sinemurian (Lucas and Tanner 2004; Gretz et al. 2015; Lathuilière and Marchal 2009). In fact that gap is not so important because the coral diversity in the Hettangian is already important (Kiessling et al. 2009). The oldest known lower Jurassic reef is located in the Hettangian from Ardèche (southern France) and was called Elmi's reef (Kiessling et al. 2009). It was discovered by Elmi and Mouterde (1965) and studied in several subsequent works, summarized by Gretz et al. (2015). Some of the Elmi's corals belong to the family Zardinophyllidae (Pachythecaliina) and

© Springer International Publishing Switzerland 2016
J. Guex, *Retrograde Evolution During Major Extinction Crises*, SpringerBriefs in Evolutionary Biology, DOI 10.1007/978-3-319-27917-6_6

represent an enigmatic group belonging to the oldest Mesozoic record of stony corals. The overall architecture of the corallite has some features of the Paleozoic rugose corals but has also the recent scleractinians aragonitic mineralogy. Their new genus *Cryptosepta*, collected in the Hettangian from Ardèche has poorly developed (cryptic) septa, which is a peculiarity that extends the boundaries used to distinguish post-Palaeozoic corals and an oversimplification that strongly suggests a prominent example of proteromorphosis (Gretz et al. 2015). It might be possibly related with the Sinemurian genus Pachysmilia, a missing link to Jurassic Amphiastreidae (Fig. 6.1).

Fig. 6.1 (**a**)
Zardinophyllum (Carn) (**b**)
Pachydendron (Carn) (**c**)
Pachythecalis (Nor) (**d**)
Cryptoseptaa (Hettangian),
(**e**) *Pachysmilia* (Sin) (data
from Gretz et al. 2015)

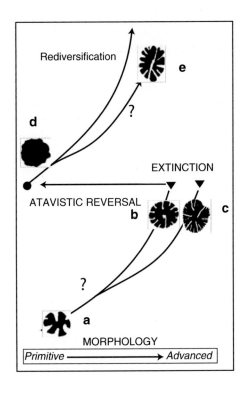

Chapter 7
Retrograde Evolution of Early Triassic Conodonts

7.1 The Conodont Animal

Conodonts were eel-shaped jawless animals, possibly related to the modern cyclostome (Goudemand et al. 2011) but their relation to the vertebrates remains under debate. They proliferated during the Paleozoic and the Triassic and disappeared during the end Triassic extinction (ETE). They were extensively studied because they evolved very fast and represent very good stratigraphic markers, sometimes better than ammonites because one finds them in many environments where ammonites are absent. These organisms have complex apatitic teeth which are generally well preserved in a variety of sediments.

7.2 Iterative Evolution in Conodonts

Triassic conodonts frequently show iterative evolutions, with repetition of heterochronous homeomorph lineages (Hirsch 1994). Specific examples of such evolutionary repetition, accompanied by a typical geometrical simplification during the transition between the ancestors and descendants, can be found above the Permian Triassic major mass extinction: the repetition of the morphogroup *Neospathodus* at the base of two successive lineages, *Clarkina* during the upper Permian and *Kashmirella-Paragondolella* during the Lower and Middle Triassic (Fig. 7.1). It is tempting to surmise that the catastrophic event, which prompted proteromorphosis coincides with the Dienerian negative carbon excursion and sea level high-stand coeval with high temperature.

© Springer International Publishing Switzerland 2016
J. Guex, *Retrograde Evolution During Major Extinction Crises*, SpringerBriefs
in Evolutionary Biology, DOI 10.1007/978-3-319-27917-6_7

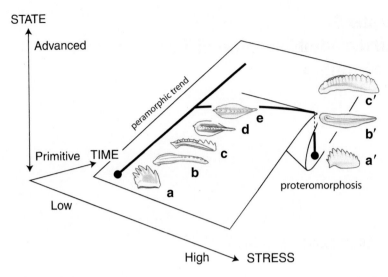

Fig. 7.1 Permian: (**a**) *"Neospathodus" arcucristatus*, (**b**) *Protoclarkina crofti*, (**c**) *Clarkina bitteri*, (**d**) *C. meishanensis*, (**e**) *C. krystyni*. Early Triassic: catastrophic reappearance of atavistic *Neospathodus kummeli* (**a′**) giving rise *to Kashmirella timorensis*, (**b′**) *Paragondolella regale* and (**c′**) *P. excelsa* (From Kilic et al. 2015)

Chapter 8
Genetic Memory in the Evolution of Silicoflagellids

8.1 Introduction

Atavisms appearing during massive extinctions or sublethal environmental stresses indicate that retrograde evolution is related to some genetic memory of the phylogenetical history of the group, recalling an Haeckelian reverse recapitulation. This clearly implies that biological experiments could be performed to reconstruct partial phylogenies of recent animals or plants by submitting them to sublethal artificial stress. Examples of such experiments are exceedingly rare; a spectacular experiment has been however conducted on living Silicoflagellids (Chrysophytes) by Van Valkenburg and Norris in 1970. This experiment demonstrates the influence of environmental stress on the development of a polymorphism oriented towards the reappearance of ancestral geometries (=atavisms).

Before describing this experiment we should recall that the development of Silicoflagellids is highly sensitive to temperature, salinity variations and pollution. These organisms are also known for their great variability which is, in part, controlled by environmental factors (Frenguelli 1935; Deflandre 1932, 1950; Van Valkenburg and Norris 1970; Dumitrica 1972; McCartney and Wise 1990; Guex 1993; McCartney et al. 2011).

Silicoflagellids first appear in the Cretaceous, and their evolution is characterized by (1) progressive increase in complexity of the skeleton by adding new elements (apical and basal systems) and (2) elongation of the skeleton with development of bilateral symmetry and concomitant simplification of some skeletal elements (loss of the lateral radial spines).

Before discussing the responses of this group to external environmental instabilities, we will briefly analyse the main characters of their geometrical evolution in three main lineages rooted in the spicular ancestor *Variramus*: *Dictyocha* on one side and *Vallacerta* and *Corbisema* on the other side.

© Springer International Publishing Switzerland 2016
J. Guex, *Retrograde Evolution During Major Extinction Crises*, SpringerBriefs in Evolutionary Biology, DOI 10.1007/978-3-319-27917-6_8

8.2   Phylogeny

The oldest representatives of the silicoflagellids appear during the lower Cretaceous (McCartney et al. 1990, 2011). Recent studies of the stratigraphic distribution provide a rather precise idea of the morphological transformations that characterize their evolution (McCartney et al. 2010). A simplified synthesis of the phylogeny of the principal silicoflagellids useful to our discussion is shown in Fig. 8.1.

The non-spicular forms of this group are classically described as composed of two main elements, the apical system and the basal ring surrounding it. The apical system has two distinct geometries: it can be composed of a simple spicule (Dictyocha), which becomes more complex and subdivides itself into an apical ring as in *Distephanus* or *Cannopilus-Paracannopilus*. Alternatively, the ancestral spicule develops an apical dome in *Schulzyocha* and *Vallacerta* (Fig. 8.2).

During the Lower and Upper Cretaceous, silicoflagellids were represented by four distinct groups: the primitive ancestral spicular form, *Variramus*, with its direct derivatives *Schulzyocha* and *Cornua*, and the genera *Vallacerta* and *Lyramula*. These primitive forms had a branched and thorny skeleton with principal rods laid out in a relatively irregular manner and their tips sometimes bi- or trifurcated. Except for several *Variramus* species, the orientation of the branches is organized in the same way as the apical part of *Dictyocha fibula*. The evolution of the *Dictyocha* lineage is schematically illustrated in Fig. 8.2. This diagram

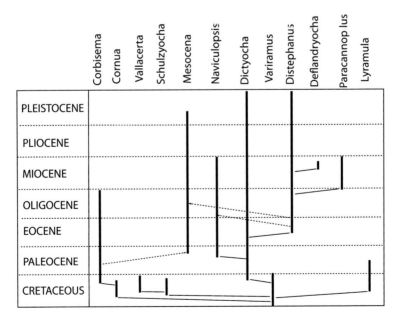

Fig. 8.1 Simplified phylogeny of the Silicoflagellids (modified from Guex 1993 and McCartney et al. 2011)

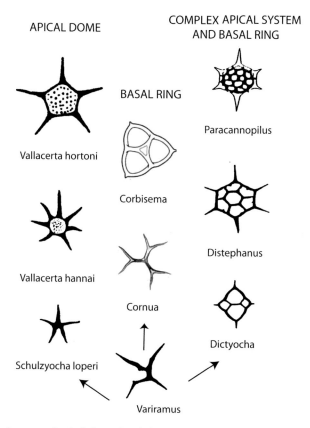

APICAL DOME

COMPLEX APICAL SYSTEM
AND BASAL RING

BASAL RING

Vallacerta hortoni

Paracannopilus

Corbisema

Vallacerta hannai

Distephanus

Cornua

Schulzyocha loperi

Dictyocha

Variramus

Fig. 8.2 Development of apical dome, basal ring and complex apical system in the three main lineages of Silicoflagellids (discussion in Guex 1993)

shows how the geometry of this group increases in complexity. Starting form the spicular *Variramus*, we observe the development of a basal ring in *Dictyocha*, followed by an apical ring in *Distephanus* which develops a more complex apical system in *Cannopilus* and *Paracannopilus*, simulating the morphology of some nassellarian radiolarians.

Another peculiar lineage of silicoflagellids is represented by the series *Variramus → Vallacerta* (Fig. 8.2), which is also characterized by an increasing complexity of the apical system of the skeleton. The trend starts with a widening of the median part of the spicule in *Schulzyocha loperi* (see McCartney et al. 2010), giving rise to an initial, weakly developed apical dome which is followed by significant development of the apical dome in *Vallacerta hannai*. Increasing regularity leads to a perfect symmetry of the spicule that characterizes *Vallacerta hortoni* and *V. quadrata*, with the development of forms with five and four horns.

Another important lineage is generated from the spicular *Cornua*, which acquires a basal ring characteristic of its direct descendant *Corbisema* (Fig. 8.2).

8.3   Genetic Memory of Silicoflagellids

The evolution of the silicoflagellids is important in the light of the fundamental observations of van Valkenburg and Norris (1970) who experimented with monoclonal cultures of *Dictyocha fibula*. On the basis of a single stock, these authors studied the development of these unicellular algae in various environmental contexts (nutrient, salinity, temperature variables). On the whole, they chose randomly and examined approximately 200 skeletons generated in their cultures. In addition to *Dictyocha fibula*, they recognized a large variety of forms assignable to the morphogenera *Distephanus*, co-occurring with individuals lacking a basal ring such as *Cornua* and *Variramus*, as well as several forms having lost all symmetry. The variability of populations obtained during this experiment is close of that of *Dictyocha fibula* described by Frenguelli (1935) in the Gulf of San Matias in Patagonia. These researchers also noted that in supersaturated cultures many cells loose their skeleton (see also Jochem and Babanerd 1989; Moestrup and Thomsen 1990).

In brief this experiment shows that a monoclonal population of protists with simple skeletons, maintained in conditions which are equivalent to an artificial ecological stress, develop an important variability with most of the morphotypes existing in the phylogenetic lineage being represented, and the development of archaic types is also observed, i.e. *Variramus* by loss of the basal ring or cells without any skeleton, like the ancestors of the whole lineage (Fig. 8.3). In other words this modern group clearly has a genetic memory of the skeletal geometry of its Cretaceous ancestors and of its phylogeny (Guex 2006).

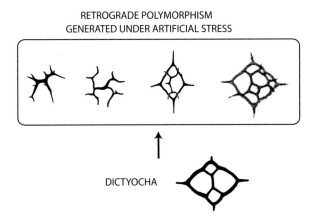

Fig. 8.3 Main morphotypes generated under artificial stress applied to a monoclonal population of *Dictyocha* in the Van Valkenburg and Norris (1970) experiment (see details in Guex 1993)

8.4 Technical Remark About the Generic Name *Distephanus*

In 1993, we noted that the widely used silicoflagellid genus *Distephanus* was pre-occupied by a terrestrial genus name of plant. As this genus was rarely used in the literature, we proposed to consider it as a *nomen oblitum* and keep the vernacular silicoflagellid name. Our solution was not followed and in 2014, Dumitrica demonstrated that *Distephanus* could be replaced by another silicoflagellid name, *Octactis*. Meanwhile, Jordan and McCartney (2015) introduced the new generic name *Stephanocha* to replace the very commonly used *Distephanus*. In the present work we chose not to get involved into the polemic and continue to use the well-known silicoflagellid *Distephanus* name, even if we entirely agree with Dumitrica's technical conclusion.

Chapter 9
Conclusions

As already noted, there are very few biological papers that propose even a partial explanation of the evolutionary reversals observed in the palaeontological data discussed above.

Several evolutionary reversions are documented in the recent literature (Hall 1984; Cabej 2011): reappearance of limbs of snakes, eyes of ostracods, mandibles in collembola, reappearance of ancestral digits of guinea pigs, horse toes, all elements that were lost during the evolution of these different zoological groups. None of these reversals seems clearly related to environmental stress and some of them have been done artificially like the teeth in chicken. The biochemical or genetical origin of these reversals is not known. Furthermore, they operate in a way strictly contrary to the cases discussed in the present paper where all the reversals concern the loss of most recent ancestral structures of the groups submitted to an extreme sublethal stress and the recovery of these structures in later evolution during more or less stable environmental conditions.

One potential source of variability and genesis of "hopeful monsters" invoked as a possible mechanism for abnormal development is a dysfunction of the heat shock protein HSP90 on the development of drosophiles (Rutherford and Lindquist 1998) under artificial chemical stress. It is clear, however, that all individuals raised in their experiment are pathological and totally dissymmetrical, contrary to most of the evolutionary cases discussed in the present paper where all the individuals keep a perfect symmetry. One case where a dysfunction of HSP90 could be invoked is the case of some completely disorganized and dissymmetrical entactinids around the Triassic–Jurassic boundary discovered by E.S. Carter (Fig. 9.1) and discussed in detail by Guex et al. (2012b).

In fact, the loss of symmetry during development has long been known as a common result of environmental stress (see, for example, the thoughtful discussion of Hoffman and Parsons, 1991). However, dysfunction of HSP90 genes is certainly not

© Springer International Publishing Switzerland 2016
J. Guex, *Retrograde Evolution During Major Extinction Crises*, SpringerBriefs
in Evolutionary Biology, DOI 10.1007/978-3-319-27917-6_9

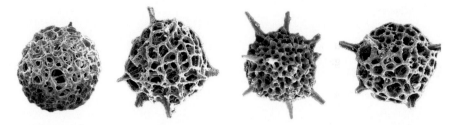

Fig. 9.1 Example of completely disorganized and dissymmetrical entactinids around the Triassic–Jurassic boundary (simplified from Guex et al. 2012b)

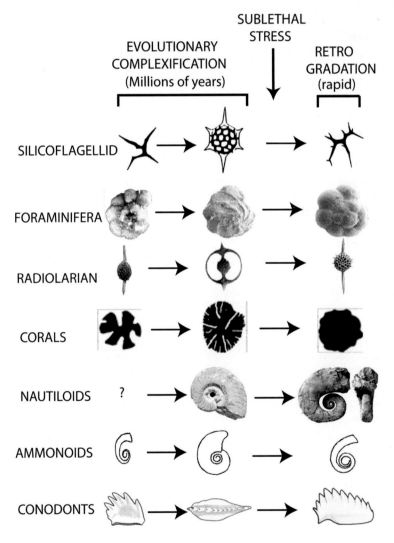

Fig. 9.2 Simplified and diagrammatic representation of the evolutionary lineages discussed in the main text and their reaction to catastrophic events. Not to scale

responsible for the appearance of the perfectly regular atavistic cephalopods and planktonic microorganisms generated during periods of major environmental perturbations.

Figure 9.2 summarizes in an extremely simplified way the evolutionary cases discussed in the main text. It shows that most of the evolutionary innovations are accumulated peramorphically (by "terminal addition" in the Haeckelian terminology) over long periods of time (Guex et al. 2012b). It is precisely these newly acquired characters which are generally lost during periods of sublethal stress that are always much shorter than the recovery periods (a few thousand years vs. several millions of years). The more generalized (=plesiomorphic) characters are also much more stable. This suggests that another possible mechanism could involve the genetic switching of Schlichtling and Pigliucci (1998) (see also Badyaev 2005). If this were true, we could speculate that some regulatory genes controlling the development of the newly acquired characters could be switched off at certain concentrations of pollutants (or under some sublethal temperature conditions) and switched on at normal concentrations. However that hypothesis cannot explain the usual cases where atavistic forms generated under extreme environmental stress are indeed giving rise to completely new lineages that are not identical to the ancestral one (compare, for example, the lineages following *Ophiceras* and *Psiloceras* in Guex (2006) (see Fig. 5.2)). Whatever the cause (genetic or biochemical) of the reversal processes described here, we note that it is very easy to inhibit the development of morphological novelties accumulated over several millions of years and to "reinitialize" the evolutionary clock of organisms submitted to high environmental perturbations leading to extinctions.

Our examples demonstrate clearly that multicellulars follow basically the same kind of evolutionary transformation (peramorphoses or atavisms) as their unicellular ancestors by accumulating new characters in an Haeckelian way (see also Torday 2015). In other words the basic morphogenetic rules are the same as in unicellulars but they are spreading all along the ontogeny of the multicellulars (Fig. 9.1).In summary we have seen that in both, unicellular and metazoans, most of the anagenetic evolution occurs by addition of new characters during long periods of time (several millions of years) (Fig. 9.1). Retrograde evolution is catastrophic in the sense of Thom (1972): as discussed in Chap. 1, we use the cusp catastrophe to represent the evolutionary retrogradations occurring during extreme environmental stress (instantaneous in geological time). These diagrams show one simple thing: we can note that the retrogradations occur in exactly the same way in protists/unicellular and in metazoans (see Figs. 5.13 and 9.1). In other words it seems that the problem can be reduced to isolated cells communicating only with the external marine environment, the morphogenesis and phylogeny being strictly controlled by intracellular signals, which are not yet well understood.

References

Adams, C. G. (1983). Speciation, phylogenesis, tectonism, climate and eustasy: Factors in the evolution of Cenozoic larger foraminiferal bioprovinces. *Systematic Association Special Publication, 23*, 255–289.

Ando, A., & Huber, B. T. (2005). *Origin of the Rotalipora montsalvensis–cushmani lineage and its implication for generic classification of the keeled Rotaliporinae.* 2nd Meeting of the Mesozoic Planktonic Foraminifera Working Group, Fribourg (pp. 25–27).

Badyaev, A. V. (2005). Role of stress in evolution: From individual adaptability to evolutionary adaptation. In B. Hallgrimson & B. K. Hall (Eds.), *Variation* (pp. 277–302). New York: Elsevier.

Barras, C., Fontanier, C., Jorissen, F., & Hohenegger, J. (2010). A comparison of spatial and temporal variability of living benthic foraminiferal faunas at 550 m depth in the Bay of Biscay. *Micropaleontology, 56*(3/4), 275–295.

Bartolini, A., & Larson, R. L. (2001). Pacific Microplate and the Pangea supercontinent in the Early to Middle Jurassic. *Geology, 29*(8), 735–738.

Bartolini, A., Guex, J., Spangenberg, J., Taylor, D., Schoene, B., Schaltegger, U., et al. (2012). Disentangling the Hettangian carbon isotope record: Implications for the aftermath of the end-Triassic mass extinction. *Geochemistry, Geophysics and Geosystems, 13*, 1–11. doi:10.1029/2011GC003807.

Bartolini, A., Nocchi, M., Baldanza, A., & Parisi, G. (1990). *Benthic life during the early Toarcian anoxic event in the Southwestern Tethyan Umbria-Marche basin, Central Italy* (pp. 323–338). Benthos'90, Sendai, Japan: Tokai University Press.

Benton, M. J. (1995). Diversification and extinction in the history of life. *Science, 5207*, 52–58.

Bijlsmal, R., & Loeschcke, V. (2005). Environmental stress, adaptation and evolution: An overview. *Journal of Evolutionary Biology, 18*(4), 744–749.

Blow, W. H. (1956). Origin and evolution of the foraminiferal genus *Orbulina* D'Orbigny. *Micropaleontology, 2*, 57–70.

Bolli, H. M., & Saunders, J. B. (1985). Oligocene to Holocene low latitude planktic foraminifers. In H. M. Bolli, J. B. Saunders, & K. Perch-Nielsen (Eds.), *Plankton stratigraphy* (pp. 155–262). Cambridge, England: Cambridge University Press.

Brönnimann, P., & Brown, N. K. (1955). Taxonomy of Globotruncanidae. *Eclogae Geologicae Helvetiae, 48*, 503–561.

Buckman, S. S. (1887–1907). Monograph of the ammonites of the Inferior Oolite Series. *Paleontology Society (London), 262*, 456 p.

Cabej, N. (2011). *Epigenetic principles of evolution.* London: Elsevier. 864 p.

Camacho, S. G., de Jesus Moura, D. M., Connor, S., Scott, D. B., & Boski, T. (2015). Taxonomy, ecology and biogeographical trends of dominant benthic foraminifera species from an

Atlantic-Mediterranean estuary (the Guadiana, southeast Portugal). *Palaeontologia Electronica,* *18*(1), 17A.

Cardenas, A. L., & Harries, P. J. (2010). Effect of nutrient availability on marine origination rates throughout the Phanerozoic eon. *Nature Geoscience, 3,* 430–434.

Caron, M. (1985). Cretaceous planktic foraminifera. In H. M. Bolli, J. B. Saunders, & K. Perch-Nielsen (Eds.), *Plankton stratigraphy* (pp. 17–86). Cambridge, England: Cambridge University Press.

Carter, E. S. (1993). Biochronology and paleontology of uppermost Triassic (Rhaetian) radiolarians, Queen Charlotte Islands, British Columbia, Canada. *Mémoires de Géologie (Lausanne), 11,* 1–175.

Carter, E. S., Gorican, Š., Guex, J., O'Dogherty, L., De Wever, P., Dumitrica, P., et al. (2010). Global radiolarian zonation for the Pliensbachian, Toarcian and Aalenian. *Palaeogeography Palaeoclimatology Palaeoecology, 297,* 401–419.

Carter, E. S., & Guex, J. (1999). Phyletic trends in uppermost Triassic (Rhaetian) Radiolaria: Two examples from Queen Charlotte Islands, British Columbia, Canada. *Micropaleontology, 45,* 183–200.

Cifelli, R. (1969). Radiation of Cenozoic planktonic foraminifera. *Systematic Zoology, 18,* 154–168.

Clemence, M., Gardin, S., Bartolini, A., Paris, G., Beaumont, V., & Guex, J. (2010). Bentho-planktonic evidence from the Austrian Alps for a decline in sea-surface carbonate production at the end of the Triassic. *Swiss Journal of Geosciences, 103,* 293–315.

Cohen, A. S., & Coe, A. L. (2007). The impact of the Central Atlantic Magmatic Province on climate and on the Sr- and Os-isotope evolution of seawater. *Palaeogeography, Palaeoclimatology, Palaeoecology, 244,* 374–390.

Cope, E. D. (1896). *The primary factors of organic evolution.* Chicago: Open Court Publishing Company. 547 pp.

Courtillot, V. (1999). *Evolutionary catastrophes: The science of mass extinction.* Cambridge, England: Cambridge University Press. 173 p.

Courtillot, V., & Gaudemer, Y. (1996). Effects of mass extinctions on biodiversity. *Nature (London), 381,* 146–148.

De Wever, P., Dumitrica, P., Caulet, J. P., Nigrini, C., & Caridroit, M. (2001). *Radiolarians in the sedimentary record.* London: Gordon and Breach Science. 553 p.

De Wever, P., O'Dogherty, L., & Gorican, S. (2006). The plankton turnover at the Permo-Triassic boundary, emphasis on radiolarians. *Eclogae geologicae Helvetiae, 99*(Suppl. 1), S49–S62.

Deflandre, G. (1932). Sur la systématique des Silicoflagellés. *Bulletin de la Société Botanique de France, 79,* 494–506.

Deflandre, G. (1950). Contribution à l'étude des Silicoflagellidés actuels et fossiles. *Microscopie, 2,* 1–82.

Dera, G., Neige, P., Dommergues, J.-L., Fara, E., Laffont, R., & Pellenard, P. (2010). High-resolution dynamics of Early Jurassic marine extinctions: The case of Pliensbachian–Toarcian ammonites (Cephalopoda). *Journal of the Geological Society, 167,* 21–33.

Desmares, D., Grosheny, D., & Beaudoin, B. (2003). Hétérochronies du développement sensu Gould chez les foraminifères planctoniques cénomaniens: Exemple de néoténie dans le bassin du Western Interior américain. *Comptes Rendus Palevol, 2,* 587–595.

Desmares, D., Grosheny, D., & Beaudoin, B. (2008). Ontogeny and phylogeny of Upper Cenomanian rotaliporids (Foraminifera). *Marine Micropaleontology, 69,* 91–105.

Douady, S., & Couder, Y. (1992). Phyllotaxis as a physical self-organized growth process. *Physical Review Letters, 68*(13), 2098–2101.

Dumitrica, P. (1970). Cryptocephalic and cryptothoracic Nassellaria in some Mesozoic deposits of Romania. *Revue Roumaine de Géologie Géophysique et Géographie (Série Géologie), 14,* 45–124.

Dumitrica, P. (1972). Miocene and Quaternary Silicoflagellates in sediments from the Mediterranean Sea. *Initial Reports of the Deep Sea Drilling Project, 13,* 902–933.

Dumitrica, P. (1982). Triassic Oertlisponginae (Radiolaria) from eastern Carpathians and southern Alps. *Dari de Seama ale Sedintelor—Institutul de Geologie si Geofizica, 6713,* 57–74.

Dumitrica, P. (1985). Internal morphology of the Satumalidae (Radiolaria): Systematic and phylogenetic consequences. *Review of Micropaleontology, 28*(3), 181–196.

Durham, J. W. (1966). Clypeasteroids. In R. C. Moore (Ed.), *Treatise on invertebrate paleontology* (Echinodermata, Vol. 3, pp. 450–491). Lawrence: University of Kansas Press.

Eicher, D. L. (1972). Phylogeny of the late Cenomanian planktonic foraminifer *Anaticinella multiloculata* (Morrow). *Journal of Foraminiferal Research, 2,* 184–190.

Elmi, S., & Mouterde, R. (1965). Le Lias inférieur et moyen entre Aubenas et Privas (Ardèche). *Travaux du laboratoire de géologie de la faculté des sciences de Lyon, 12,* 143–246.

Erben, H. K. (1966). Über den Ursprung der Ammonoidea. *Biological Reviews, 41,* 641–658.

Frenguelli, J. (1935). Varaciones de Dictyocha fibula en el Golfo de San Matias. *Annales del Museo Argentino de Ciencias Naturales, 38,* 265–281.

García-Joral, F., Gómez, J. J., & Goy, A. (2011). Mass extinction and recovery of the Early Toarcian (Early Jurassic) brachiopods linked to climate change in Northern and Central Spain. *Palaeogeography Palaeoclimatology Palaeoecology, 302,* 367–380.

Gargouri, S., & Vachard, D. (1988). Sur Hemigordius et d'autres foraminifères porcelanés du Murghabien du Tebaga (Tunisie). *Review of Paleobiology, Spec. 2,* 57–68.

Gómez, J. J., Goy, A., & Canales, M. L. (2008). Seawater temperature and carbon isotope variations in belemnites linked to mass extinction during the Toarcian (Early Jurassic) in Central and Northern Spain, comparison with other European sections. *Palaeogeography Palaeoclimatology Palaeoecology, 258,* 28–58.

Gonzalez-Donoso, J. M., Linares, D., & Robaszynski, F. (2007). The rotaliporids, a polyphyletic group of Albian–Cenomanian planktonic foraminifera: Emendation of genera. *Journal of Foraminiferal Research, 37,* 175–186.

Gorican, Š., Carter, E. S., Dumitrica, P., Whalen, P. A., Hori, R. S., De Wever, P., et al. (2006). *Catalogue and systematics of Pliensbachian, Toarcian and Aalenian radiolarian genera and species.* Ljubljana, Slovenia: Založba ZRC/ZRC Publishing, ZRC SAZU. 446 p.

Gorican, Š., Carter, E. S., Guex, J., O'Dogherty, L., De Wever, P., Dumitrica, P., et al. (2013). Evolutionary patterns and palaeobiogeography of Pliensbachian and Toarcian Radiolaria. *Palaeogeography Palaeoclimatology Palaeoecology, 386,* 620–636.

Goudemand, N. L., Orchard, M. J., Urdy, S., Bucher, H., & Tafforeau, P. (2011). Synchrotron-aided reconstruction of the conodont feeding apparatus and implications for the mouth of the first vertebrates. *Proceedings of the National Academy of Sciences, 108*(21), 1–5.

Gould, S. J. (1996). *Full house.* Cambridge, England: Harvard University Press. 244 p.

Gretz, M., Lathuilière, B., & Martini, R. (2015). A new coral with simplified morphology from the oldest known Hettangian (Early Jurassic) reef in southern France. *Acta Palaeontologica Polonica, 60*(2), 277–286.

Grünig A. (1984). *Phenotypic variation in Spiroplectammina, Uvigerina and Bolivina.* 2nd International Symposium on Benthic Foraminifera (Pau, pp. 249–255).

Guex, J., Schoene, B., Bartolini, A., Spangenberg, J., Schaltegger, U., O'Dogherty, L., et al. (2012a). Geochronological constraints on post-extinction recovery of the ammonoids and carbon cycle perturbations during the Early Jurassic. *Palaeogeography, Palaeoclimatology, Palaeoecology, 346–347,* 1–11.

Guex, J., Bartolini, A., Spangenberg, J., Vicente, J. C., & Schaltegger, U. (2012b). Ammonoid multi-extinction crises during the late Pliensbachian Toarcian and carbon cycle instabilities. *Solid Earth, 4,* 1205–1228. Discussion.

Guex, J., Morard, A., Bartolini, A., & Morettini, E. (2001). Découverte d'une importante lacune stratigraphique à la limite Domérien-Toarcien; implications paléo-océanographiques. *Bulletin de la Société Vaudoise des Sciences Naturelles, 87*(3), 277–284.

Guex, J., O'Dogherty, L., Carter, E. S., Gorican, S., Dumitrica, P., & Bartolini, A. (2012b). Geometrical transformations of selected Mesozoic radiolarians. *Geobios, 45,* 541–554.

Guex, J. (1967). Contribution à l'étude des blessures chez les ammonites. *Bulletin de la Société Vaudoise des Sciences Naturelles, 69,* 323–338.

Guex, J. (1975). Description biostratigraphique du Toarcien superieur de la bordure sud des Causses. *Eclogae Geologicae Helvetiae, 68*(1), 97–129.

Guex, J. (1981). Associations virtuelles et discontinuités dans la distribution des espèces fossiles. *Bulletin de la Société´ Vaudoise des Sciences Naturelles, 75*, 179–197.

Guex, J. (1992). Origine des sauts évolutifs chez les ammonites. *Bulletin de la Société´ Vaudoise des Sciences Naturelles, 82*(2), 117–144.

Guex, J. (1993). Simplifications géométriques liées au stress écologique chez certain protistes. *Bulletin de la Société´ Vaudoise des Sciences Naturelles, 82*(4), 357–368.

Guex, J. (1995). Ammonites hettangiennes de la Gabbs Valley Range (Nevada USA). *Mémoires de géologie Lausanne, 27*, 1–131.

Guex J. (1999). Taxonomy and paleobiology in ammonoids biochronology: Sexual dimorphism, covariation and septal spacing. In: J. Savary & J. Guex (Eds.) Discrete biochronological scales and unitary associations: Description of the BioGraph computer program. *Mémoires de Géologie Lausanne, 34*, 42–43.

Guex, J. (2000). Paronychoceras gen.n., un nouveau genre d'ammonites (Cephalopoda) du Lias supérieur. *Bulletin de la Société Vaudoise des Sciences Naturelles, 82*, 115–124.

Guex, J. (2001). Involution croissante et règle de Cope. *Bulletin de la Société´ Vaudoise des Sciences Naturelles, 87*, 373–379.

Guex, J. (2003). A generalization of Cope's rule. *Bulletin de la Societe Geologique de France, 174*(5), 449–452.

Guex, J. (2006). Reinitialization of evolutionary clocks during sublethal environmental stress in some invertebrates. *Earth and Planetary Science Letters, 240*, 242–253.

Guex, J., Bartolini, A., Atudorei, V., & Taylor, D. (2004). High-resolution ammonite and carbon isotope stratigraphy across the Triassic—Jurassic boundary at New York Canyon (Nevada). *Earth and Planetary Science Letters, 225*, 29–41.

Guex, J., Caridroit, M., Kuwahara, K., & O'Dogherty, L. (2014). Retrograde evolution of *Albaillella* during the Permian-Triassic crisis. *Revue de Micropaleontologie, 57*, 39–43.

Guex, J., Galster, F., & Hammer, O. (2015a). *Discrete biochronological time scales*. New York: Springer. 160 p.

Guex, J., Koch, A., O'Dogherty, L., & Bucher, H. (2003). A morphogenetic explanation of Buckmans's law of covariation. *Bulletin de la Société géologique de France, 174*, 603–606.

Guex J., Pilet S., Muntener O., Bartolini A., Spangenberg J., Schoene B., Schaltegger U. (2015). Thermal erosion of cratonic lithosphere as a trigger for mass-extinction. Manuscript submitted.

Haas, O. (1942). Recurrence of morphologic types and evolutionary cycles in Mesozoic ammonites. *Journal of Paleontology, 16*, 643–650.

Hall, B. K. (1984). Developmental mechanisms underlying the formation of atavisms. *Biological Reviews, 59*, 89–124.

Hallam, A., & Wignall, P. B. (1997). *Mass extinctions and their aftermath*. Oxford: Oxford University Press. 320 p.

Hammer, Ø., & Bucher, H. (2005). Buckman's first law of covariation, a case of proportionality. *Lethaia, 38*, 67–72.

Hammer, O., Guex, J., & Savary, J. (2015). *The Uagraph program*. Retrieved from http://folk.uio.no/ohammer/uagraph.

Harazim, D., Van de Schootbrugge B. Sorichter K., Fiebig J., Weug A. Suan G., & Oschmann W. (2012). Spatial variability of watermass conditions within the European Epicontinental Seaway during the Early Jurassic (Pliensbachian–Toarcian). *Sedimentology*. doi:10.1111/j.1365-3091.2012.01344.x

Hart, M. B. (Ed.). (1996). Biotic recovery from mass extinction events. *Geological Society of London Special Publication, 102*, 392 pp.

Hassan, M. A., Westermann, G. E. G., Hewitt, R. A., & Dokainish, M. A. (2002). Finite-element analysis of simulated ammonoid septa (extinct Cephalopoda): Septal and sutural complexities do not reduce strength. *Paleobiology, 28*, 113–126.

Hesselbo, S. P., Jenkyns, H. C., Duarte, L. V., & Oliveira, L. C. V. (2007). Carbon-isotope record of the Early Jurassic (Toarcian) Oceanic Anoxic Event from fossil wood and marine carbonate (Lusitanian Basin, Portugal). *Earth and Planetary Science Letters, 253*, 455–470.

Hirsch, F. (1994). Triassic conodont multielements versus eustatic cycles. *Mémoires de Géologie (Lausanne), 22*, 35–52.

Hangartner, S., Laurila, A., & Räsänen, K. (2011). Adaptive divergence of the moor frog (Rana arvalis) along an acidification gradient. *Evolutionary Biology, 11*, 366.Hoffman, A. A., & Parsons, P. A. (1991). *Evolutionary genetics and environmental stress*. Oxford, England: Oxford Science. 284 p.

Hottinger, L. (1981). *The resolution power of the biostratigraphic clock based on evolution and its limits*. In J. Martinell (Ed.), International Symposium on Concepts and Methods of Paleontology (pp. 233–242), Barcelona.

Hottinger, L., & Drobne, K. (1988). Alvéolines tertiaires: Quelques problèmes liés à la conception de l'espèce. *Revue de Paléobiologie, Genève, Spéc. 2*, 665–681.

Howarth, M. K. (1973). The stratigraphy and ammonite fauna of the upper Liassic grey shales of the Yorkshire coast. *Bulletin of the British Museum of Natural History, 24*(3), 237–277.

Hudson, N. (2003). Stratigraphy and correlation of the Ururoan and Temaikan Stage (Lower-Middle Jurassic) sequences, New Zealand. *Journal of the Royal Society of New Zealand, 33*, 1. doi:10.1080/03014223.2003.9517724.

Hürtzeler, J. (1962). Kann die biologische Evolution wie sie sich in die Vergangenheit abgespielt hat, exakt erfasst werden? *Schöpfungsglaube und biologische Entwicklungslehre., 16*, 15–36.

Hyatt, A. (1869). Genesis of the arietitidae. *Museum of Comparative Zoology, Harvard, Memoir, 16*, 1–238.

Jablonka, E. (2013). Epigenetic inheritance and plasticity: The responsive germline. *Progress in Biophysics and Molecular Biology, 111*, 99–107.

Jablonka, E., & Raz, G. (2009). Transgenerational epigenetic inheritance: Prevalence, mechanisms, and implications for the study of heredity and evolution. *The Quarterly Review of Biology, 84*, 31–76.

Jenkyns, H. C. (1988). The Early Toarcian (Jurassic) anoxic event. Stratigraphic, sedimentary and geochemical evidence. *American Journal of Science, 288*, 101–151.

Jochem, F., & Babanerd, B. (1989). Naked Dictyocha speculum—a new type of phytoplankton bloom in the Western Baltic. *Marine Biology, 103*, 373–379.

Jones, C. J., Jenkyns, H. C., Coe, A. L., & Hesselbo, S. P. (1994). Strontium isotopic variations in the Jurassic Cretaceous seawater. *Geochimica Cosmochimica Acta, 58*, 3061–3074.

Jordan, R. W., & McCartney, K. (2015). *Stephanocha* nom. nov., a replacement name for the illegitimate silicoflagellate genus *Distephanus* (Dictyochophyceae). *Phytotaxa, 201*(3), 177–187.

Keller, G., & Pardo, A. (2004). Age and paleoenvironment of the Cenomanian—Turonian global stratotype section and point at Pueblo, Colorado. *Marine Micropaleontology, 51*, 95–128.

Kennett, J. P., & Srinivasan, M. S. (1983). *Neogene planktonic foraminifera*. Stroudsburg, PA: Hutchinson Ross.

Kiessling, W., Roniewicz, E., Villier, L., Léonide, P., & Struck, U. (2009). An early Hettangian coral reef in southern France: Implications for the end-Triassic reef crisis. *Palaios, 24*, 657–671.

Kilic, A.M., Plasencia, P, Ishida, K., Guex, J., Hirsch F (2015) Proteromorphosis of Neospathodus (Conodonta) during the Permian-Triassic crisis. Revue de Micropaléontologie. In press

Kishony, R., & Leibler, S. (2003). Environmental stresses can alleviate the average deleterious effect of mutations. *Journal of Biology, 2*, 14.

Korte, C., Hesselbo, S. P., Jenkyns, H. C., Rickaby, R. E. M., & Spotl, C. (2009). Palaeoenvironmental significance of carbon- and oxygen-isotope stratigraphy of marine Triassic–Jurassic boundary sections in SW Britain. *Journal of the Geological Society, 166*, 431–445. doi:10.1144/0016-76492007-177.

Kozur, H., & Mostler, H. (1983). The polyphyletic origin and the classification of the Mesozoic saturnalids (Radiolaria). *Geol. Paleont. Mitt. Innsbruck, 13*, 1–47.

Kozur, H., & Mostler, H. (1990). Saturnaliacea Deflandre and sorne other stratigraphically important Radiolaria from the Hettangian of Lenggries/Isar (Bavaria, northern Calcareous Alps). *Geol. Paleont. Mitt. Innsbruck, 17*, 179–248.

Kuwahara, K. (1999). Phylogenetic Lineage of Late Permian *Albaillella* (Albaillellaria, Radiolaria). *Journal of Geosciences. Osaka City University, 42*, 85–101.

Lathuilière, B., & Marchal, D. (2009). Extinction, survival and recovery of corals from the Triassic to Middle Jurassic time. *Terra Nova, 21*, 57–66.

Leckie, R. M. (1985). Foraminifera of the Cenomanian–Turonian boundary interval, Greenhorn Formation, Rock Canyon Anticline, Pueblo, Colorado. In L. M. Pratt, E. G. Kauffman, & F. B. Zelt (Eds.), *Fine-grained deposits of cyclic sedimentary processes* (pp. 139–149). Golden, CO: SEPM Field Trip Guidebook.

Less, G., & Kovacs, L. O. (1996). Age-estimates by European Paleogene Orthophragminas using numerical evolutionary correlation. *Geobios, 29*, 261-185.

Lloyd, S. (2001). Measures of complexity: A nonexhaustive list. *Control Systems IEEE, 21*(4), 7–8.

Longoria, J. F. (1973). Pseudoticinella, a new genus of planktonic foraminifera from the early Turonian of Texas. *Rev. esp. Micropaleont., 5*, 417–423.

Lucas, S. G., & Tanner, L. H. (2004). Late Triassic extinction events. *Albertiana, 31*, 31–40.

Magaritz, M. (1989). δ13C minima follow extinction events: A clue to faunal radiation. *Geology, 17*(4), 337–340.

Mancini, E. A. (1978). Origin of micromorph faunas in the geologic record. *Journal of Paleontology, 52*(2), 311–322.

Marjanac, T., & Steel, R. J. (1997). Dunlin Group sequence stratigraphy in the northern North Sea A model for Cook sandstone deposition. *AAPG Bulletin, 81*(2), 276–292.

McCartney, K., & Wise, S. W. (1990). Cenozoic Silicoflagellates and Ebridians from ODP leg 113: Biostratigraphy and notes on morphologie variability. *Proceeding of ODP Scientific Results, 113*, 729–760.

McCartney, K., Wise, S. W., Harwooo, D. M., & Gersonde, R. (1990). Enigmatic Lower Albian Silicoflagellates from ODP site 693: Progenitors of the order Silicoflagellata? *Proceeding of ODP Scientific Results, 113*, 427–442.

McCartney, K., Witkowski, J., & Harwood, D. M. (2010). Early evolution of the silicoflagellates during the Cretaceous. *Marine Micropaleontology, 77*, 83–100.

McCartney, K., Witkowski, J., & Harwood, D. M. (2011). Unusual assemblages of Late Cretaceous silicoflagellates from the Canadian Archipelago. *Revue de Micropaléontologie, 54*, 31–58.

McElwain, J. C., Beerling, D. J., & Woodward, F. I. (1999). Fossil plants and global warming at the Triassic-Jurassic Boundary. *Science, 285*, 1386–1390.

McElwain, J. C., Wagner, P. J., & Hesselbo, S. P. (2009). Fossil plant relative abundances indicate sudden loss of Late Triassic biodiversity in East Greenland. *Science, 324*, 1554–1556. doi:10.1126/science.1171706.

McKinney, M. L. (1990). Trends in body-size evolution. In K. J. McNamara (Ed.), *Evolutionary trends* (pp. 75–118). London: Belhaven.

McShea, D. W. (1991). Complexity and evolution: What everybody knows. *Biology and Philosophy, 6*, 303–324.

Meinhardt, H., Koch, A. J., & Bernasconi, G. (1998). Models of pattern formation applied to plant development. In D. Barabe & R. V. Jean (Eds.), *Symmetry in plants* (pp. 723–758). Singapore: World Scientific.

Meister, C. (1988). Ontogenèse et évolution des Amaltheidae (Ammonoidea). *Eclogae Geologicae Helvetiae, 81*(3), 763–841.

Meister, C. (1989). *Les ammonites du Domérien des Causses (France): Analyses paléontologiques et stratigraphiques* (pp. 1–80). Paris: Cahiers de Paléontologie du CNRS.

Meister, C., & Stampfli, G. (2000). Les ammonites du Lias moyen (Pliensbachien) de la Néotéthys et de ses confins; compositions fauniques, affinités paléogéographiques et biodiversité. *Revue de Paléobiologie, 19*, 227–292.

Mercker, M., Hartmann, D., & Marciniak-Czochra, A. (2013). A mechanochemical model for embryonic pattern formation: Coupling tissue mechanics and morphogen expression. *PLoS One, 8*(12), e82617.

Moestrup, O., & Thomsen, H. A. (1990). *Dictyocha speculum* (Silicoflagellate, Dictyochaphyceae) studies on armoured and non armoured stages. *Biologiske Skrifter, 37*, 1–56.

Monnet, C., Bucher, H., Guex, J., & Wasmer, M. (2012a). Large-scale evolutionary trends of Acrochordiceratidae Arthaber, 1911 (Ammonoidea, Middle Triassic) and Cope's Rule. *Palaeontology, 55*(1), 87–107.

Monnet, C., Bucher, H., & Brayard, A. (2012b). Globacrochordiceras gen. nov. (Acrochordiceratidae, late Early Triassic) and its significance for stress-induced evolutionary jumps in ammonoid lineages (cephalopods). *Journal of Systematic Palaeontology, 16*, 197–215.

Morard, A., & Guex, J. (2003). Ontogeny and covariation in the Toarcian genus Osperleioceras (Ammonoidea). *Bulletin de la Société Géologique de France, 174*(6), 607–615.

Muller, S. W., & Ferguson, H. G. (1939). Mesozoic stratigraphy of the Hawthorne and Tonopah quadrangles, Nevada. *Geological Society of America Bulletin, 50*, 1573–1624.

Newell, A. C., Shipman, P. D., & Sun, Z. (2008). Phyllotaxis: Cooperation and competition between mechanical and biochemical processes. *Journal of Theoretical Biology, 251*, 421–439.

Nevo, E. (2011). Evolution under environmental stress at macro- and microscales. *Genome Biology and Evolution, 3*, 1039–1052.

O'Dogherty, L., Sandoval, J., Bartolini, A., Bill, M., Bruchez, S., & Guex, J. (2006). Carbon-isotope stratigraphy and ammonite faunal turnover for the Middle Jurassic in the southern Iberian paleomargin. *Palaeogeography Palaeoclimatology Palaeoecology, 239*, 311–333.

Pálfy, J., & Smith, P. L. (2000). Synchrony between Early Jurassic extinction, oceanic anoxic event, 560 and the Karoo–Ferrar flood basalt volcanism. *Geology, 28*, 747–750.

Peybernes, B., Fondecave-Wallez, M. J., Gourinard, Y., & Eichene, P. (1997). Comparative sequence stratigraphy and grade-dating by Planktonic Foraminifera from the Campanian–Maestrichtian and Paleocene of Several Sections in Southwestern Europe and North Africa. *Comptes-Rendus de l'Academie des Sciences, Paris, 324*, 839–846.

Price, G. D. (1999). The evidence and implications of polar ice during the Mesozoic. *Earth-Science Reviews, 48*(3), 183–210.

Raff, R. A., & Kaufman, T. C. (1983). *Embryos, genes, and evolution.* New York: Macmillan. 395 p.

Raup, D. (1967). Geometric analysis of shell coiling: Coiling in ammonids. *Journal of Paleontology, 41*, 43–65.

Riedel, W. R., & Sanfilippo, A. (1981). Evolution and diversity of form in Radiolaria. In T. L. Simpson & B. E. Volcani (Eds.), *Silicon and siliceous structures in biological systems* (pp. 323–346). New York: Springer.

Robaszynski, F., & Caron, M. (1995). Foraminifères planctoniques du Crétacé: Commentaire de la zonation Europe-Méditerranée. *Bulletin de la Societe Geologique de France, 166*, 681–692.

Runnegar, B. (1987). Subphylum cyrtostoma, class monoplacophora. In R. S. Boardman, A. H. Cheetham, & A. J. Rowell (Eds.), *Fossil invertebrates* (pp. 297–304). Palo Alto, CA: Blackwell.

Rutherford, S. L., & Lindquist, S. (1998). Hsp90 as a capacitator for morphological evolution. *Nature (London), 396*, 336–342.

Sanfilippo, A., & Riedel, W. R. (1970). Post-Eocene "closed" theoperid radiolarians. *Micropaleontology, 16*(4), 446–462.

Sanfilippo, A., & Riedel, W. R. (1982). Revision of the radiolarian genera Theocotyle, Theocotylissa and Thyrsocyrtis. *Micropaleontology, 28*, 170–188.

Sanfilippo, A., Westberg-Smith, M. J., & Riedel, W. R. (1985). Cenozoic Radiolaria. In H. M. Bolli, J. B. Saunders, & K. Perch-Nielsen (Eds.), *Plankton stratigraphy* (pp. 631–712). Cambridge, England: Cambridge University Press.

Schaller, M. F., Wright, J. D., & Kent, D. V. (2011). Atmospheric PCO_2 perturbations associated with the Central Atlantic Magmatic Province. *Science, 331*, 1404–1409. doi:10.1126/1199011.

Schindewolf, O. H. (1940). "Konvergenzen" bei Korallen und bei Ammoneen. *Fortschritte in Geologie und Paläontologie, 12*(41), 387–491.

Schlichtling, C. D., & Pigliucci, M. (1998). *Phenotypic evolution: A reaction norm perspective* (p. 340). Sunderland, MA: Sinauer Associates.

Schmidt, D., Thierstein, H. R., Bollmann, J., & Schiebel, R. (2004). Abiotic forcing of plankton evolution in the Cenozoic. *Science, 303*, 207–210.

Schoene, B., Guex, J., Bartolini, A., Schaltegger, U., & Blackburn, T. J. (2010). Correlating the end-Triassic mass extinction and flood basalt volcanism at the 100 ka level. *Geology, 38*, 387–390. doi:10.1130/G30683.1.

Sell, B., Ovtcharova, M., Guex, J., Bartolini, A., Jourdand, F., Spangenberg, J. E., et al. (2014). Evaluating the temporal link between the Karoo LIP and climatic–biologic events of the Toarcian Stage with high-precision U–Pb geochronology. *Earth and Planetary Science Letters, 408*, 48–56.

Sepkoski, J. J. (1978). A kinetic model of Phanerozoic taxonomic diversity: I. Analysis of marine orders. *Paleobiology, 4*(3), 223–251.

Septfontaine, M. (1988). Vers une classification évolutive des Lituolidés (Foraminifères) jurassiques en milieu de plate-forme carbonatée. *Revue de Paléobiologie, Genève, Spéc. no 2*, 229–256.

Shigeta, Y. (2009). Recovery of nautiloids in the Early Triassic. *National Museum of Nature and Science Monographs, 38*, 40–43.

Shimer, H. W. (1908). Dwarf faunas. *The American Naturalist, 42*, 472–490.

Sobolev, E. S. (1994). Stratigraphic range of Triassic boreal Nautiloid. *Mémoires de géologie Lausanne, 22*, 127–138.

Sobolev, S. V., Sobolev, A. V., Kuzmin, D. V., Krivolutskaya, N. A., Petrunin, A. G., & Arndt, N. T. (2011). Linking mantle plumes, large igneous provinces and environmental catastrophes. *Nature, 477*, 312–316. doi:10.1038/nature10385.

Spath, L. F. (1933). The evolution of the Cephalopoda. *Biological Reviews, 8*(4), 418–462.

Stanley, S. M. (1973). An explanation for Cope's rule. *Evolution, 27*(1), 1–26.

Stanley, G. D. (2003). The evolution of modern corals and their early history. *Earth-Science Reviews, 60*, 195–225.

Suan, G., Nikitenko, B. L., Rogov, M. A., Baudin, F., Spangenberg, J. E., Knyazev, V. G., et al. (2011). Polar record of Early Jurassic massive carbon injection. *Earth and Planetary Science Letters, 312*, 102–113.

Svensen, H., Planke, S., Polozov, A. G., Schmidbauer, N., Corfu, F., Podladchikov, Y., et al. (2009). Siberian gas venting and the end-Permian environmental crisis. *Earth and Planetary Science Letters, 277*, 490–500.

Takemura, A., & Aono, R. (2007). Systematic Description of some radiolarians from Induan cherts in the ARH and ARF sections of the Oruatemanu formation, Arrow Rocks, New Zealand. In K. B. Spörli, A. Takemura, & R. S. Hori (Eds.), *The oceanic Permian/Triassic boundary sequence at Arrow Rocks (Oruatemanu), Northland, New Zealand* (GNS Science Monograph, Vol. 24, pp. 197–205). Lower Hutt, New Zealand: GNS Science.

Thom, R. (1972). *Stabilité structurelle et morphogenèse.* New York: Benjamin. 358 p.

Torday, J. S. (2015). The cell as the mechanistic basis for evolution. *Wiley Interdisciplinary Reviews: Systems Biology and Medicine, 7*(5), 275–284. doi:10.1002/wsbm.1305.

van Iterson, G. (1907). *Mathematische und microscopisch-anatomische Studien ueber Blattstellungen, nebst Betraschungen ueber den Schalenbau der Miloiolinen.* Jena, Germany: Gustav-Fischer.

Van Valkenburg, S. D., & Norris, R. E. (1970). The growth and morphology of the Silicoflagellate Dictyocha fibula Ehrenberg in culture. *Journal of Phycology, 6*, 48–54.

Westermann, G. E. G. (1966). Covariation and taxonomy of the Jurassic ammonite Sonninia adicra (Waagen). *Neues Jahrbuch für Geologie und Paläontologie (Abhandlungen), 124*, 289–312.

Zachos, J., Pagani, M., Ioan, L., Thomas, E., & Billups, K. (2001). Trends, Rhythms, and Aberrations in Global Climate 65 Ma to Present. *Science, 292*, 686–693.

Index

© Springer International Publishing Switzerland 2016
J. Guex, *Retrograde Evolution During Major Extinction Crises*, SpringerBriefs
in Evolutionary Biology, DOI 10.1007/978-3-319-27917-6

Printed in the United States
By Bookmasters